An Introduction to

Landscape Design and Construction

The Landscape Institute is the chartered professional body for landscape architects. The Institute embraces landscape designers, managers, scientists and landscape planning. Its objects are to protect, conserve and enhance the natural and built environment by promoting the arts and sciences of landscape architecture. The promotion of high standards of education, qualification and competence of those who practise landscape architecture are key functions of the Institute and we welcome the publication of this book *An Introduction to Landscape Design and Construction*.

Landscape architecture offers a fascinating career and provides satisfaction for those who practise the profession and those who benefit from its work. We hope this book will enlighten you and increase your desire to pursue this interesting career.

The Landscape Institute, 6–8 Barnard Mews, London SW11 1QU
Tel: 0171 738 9166; Fax: 0171 738 9134; e-mail: mail@l-l.org.uk

An Introduction to
Landscape Design and Construction

JAMES BLAKE

Gower

Published by
Gower Publishing Limited
Gower House
Croft Road
Aldershot
Hampshire GU11 3HR
England

Gower
Old Post Road
Brookfield
Vermont 05036
USA

James Blake has asserted his right under the Copyright, Designs and Patents Act 1988 to be identified as the author of this work.

British Library Cataloguing in Publication Data
Blake, James
    An introduction to landscape design and construction
    1. Landscape design
    I. Title
    712

ISBN 0–566–07769–8 Hardback
ISBN 0–566–07775–2 Paperback

Library of Congress Cataloging-in-Publication Data
Blake, Bartholomew James, 1961–
    An introduction to landscape design and construction / Bartholomew James Blake.
        p.   cm.
    Includes index.
    ISBN 0–566–07769–8 (hardcover). — ISBN 0–566–07775–2 (pbk.)
    1. Landscape design.   2. Landscape architecture.   3. Building.
    I. Title.
    SB472.45.B58   1998
    712—dc21   98–9674
                                                           CIP

Typeset in Plantin by Wileman Design, Farnham
and printed in Singapore by Kyodo Printing Co (S'pore) Pte Ltd.

Contents

List of figures

Preface

THIS book fills a gap hitherto surprisingly neglected by landscape professionals: it aims to provide a simple, direct and concise introduction to what can appear to be a vast and daunting subject, especially so to those on an introductory course or for new practitioners continuing their professional development. It is because landscape design is such an immense subject that this book cannot claim to be a complete guide and the space available for each section is limited. Its purpose then is simply to provide an overview of the subject, making the book of practical value for reference purposes – particularly for providing simple and clear definitions of the core subject of design and construction as it is practised.

My central aim has been to provide a text prepared from the viewpoint of a working practitioner, rather than from an academic perspective, and as such I hope that it will be of assistance to those about to embark on a career in landscape design as well as to the student struggling to sort out the theory from the practice. Where subjects are covered only briefly, I consider that they require separate study. Specialist books are available on these areas, which include landscape and garden history, landscape planning, environmental impact assessment, wildlife conservation and the law relating to land and landscape.

James Blake
Lavenham, Suffolk

Acknowledgements

TO the stafff at James Blake Associates: Giles Hill BA (Hons) BLA MLI – for his support and for covering my time lost to the business in writing this book; Julie Day – for her commitment and support in much editing, checking and word processing; Geoff Yates of Sunshine Surveys, Chedburgh, Nr Bury St Edmunds – for writing the surveying section of the book; Suzie Duke – for editing; Steve Coghill, Head of Horticulture at Otley College – for inspiring me to write the book in the first place and for his continued support; Andrew Stephens – for proofreading; John Blythe of 13 Station Road, Woodbridge – for the majority of the photographic illustrations; Adrian Clarke – for some of the latter photographic illustrations; Justin Kibble – draughtsman for most of the construction details; Ian McFaddian, former student at Otley College, for allowing his college project set by the author to be used for illustration of some working drawings; Mr Mike Hyder of Hy-tex UK Ltd for kind assistance with developing wheat-free planting specifications, and support for illustrations and drawings; my clients Patrick and Margaret McKenna for allowing me to photograph their private grounds for some of the illustrations. Finally, and most importantly, to my wife Louise for suffering my 'creative' tantrums with such good humour.

Introduction

AN introduction to landscape design can never hope to be a full account of every aspect of the subject, as then, by its very nature, it would no longer be an introduction but instead a series of rather laborious, dry and crusty volumes, the sort that end up as shelf fillers. An introduction should be a vehicle that provides a fast route into the subject, painting the broader picture and providing a framework of knowledge. By understanding this framework, the reader will hopefully be better placed to assimilate, interpret and use effectively any new detailed information encountered. This book is therefore like a large-scale map, a location plan, portraying how the subject fits together as a whole, providing a sound working knowledge of the basics.

It is not possible to provide all the detail about every aspect of landscape design, not least because the subject is so vast and has many related disciplines, including landscape management and landscape science. There are already books that explore the detail of various aspects of the subject, but such detailed studies are arguably less valuable to the new practitioner or student without referral to the wider picture. When faced with such an onslaught of information as landscape design encompasses, it is necessary to understand quickly the relative importance of that information. Designers encounter or research new information continuously, and evaluation of how often it will be needed in practice and how much or how little such information affects other data is essential to prevent information overload.

The value, therefore, of a text that sets out the basics in a straightforward and concise way is clear, because it will provide a rationale to enable the evaluation and understanding of all subsequent information more immediately. It is intended that by reading this book a student or new practitioner will be able to practise more effectively, with greater confidence, certain judgement and greater pragmatism. The self-confidence afforded by a sound grounding in the core subject will empower the new practitioner or student and through confident application of the basics there will be much greater opportunity for new talent to be expressed, new ideas to be sounded and much greater chance for innovative design solutions to be realized. There is no greater chance for new ideas and creative flair than from empowering new, free-thinking practitioners, possessing, as many do, great enthusiasm and energy.

CHAPTER 1

The role of the landscape designer

LANDSCAPE design is an unusual profession, mixing as it does both the arts and the sciences. Landscape designers are holistic, working with both poles of human culture, the practical and analytical as well as the artistic and innovative. The potential for the subject as an art form (with all the creative inspiration and thinking that is therefore implied) is at least as great as for fine art. After all, landscape is an art form brought to the people – the people who live and work in and around landscapes and cityscapes – whereas in the cases of music, painting and writing people need to visit a gallery, play a tape or read a book to experience the art.

Landscape designers can have an immense impact on our lives, improving the richness and beauty of our surroundings. Landscape design demands a high degree of technical proficiency, but unlike other art forms the landscape must further perform in terms of its engineering and architectural specification. The landscape must endure in perpetuity and yet also evolve with the maturity of its living 'soft' materials.

The coming generation of landscape designers will have, more than ever before, the opportunity to use their creative flair. Yet greater artistic freedom correspondingly demands greater practicality, pragmatism and precision. Even today, the growing emphasis on environmental protection means that landscape issues are increasingly high on the political agenda. As land becomes ever more scarce, it must be better and more efficiently used and better designed. The scarcity of available land will doubtless raise the financial stakes, and may provide ever greater funds with which to realize creative dreams.

Landscape design is an all-encompassing term which ranges from keen gardeners planning new flower beds in their gardens, to a million pound civil engineering and soft landscape design involving a multitude of land uses from urban street design, domestic and commercial car parking and service areas to parks, play areas and formal gardens. Clearly, while the keen gardener may well have no qualifications but sufficient experience to arrange a collection of garden plants, the gardener is very unlikely to be competent at designing the million pound civil engineering and landscape design mentioned above. Therefore, it is important to make a distinction within the various levels of landscape practitioner and most such practitioners, apart from keen gardeners, would fall between the following two categories:

1. Garden designers with some experience and qualifications, which might include National Certificates, National Diplomas and Higher National Diplomas in horticulture or garden design.
2. Landscape architects who have undergone several years of intensive full-time higher education, including degree courses and postgraduate diploma and degree courses, with the benefit of integral 'year out' employment experience. Such practitioners would go on to take professional exams following two further years of post-educational employment experience and these professional exams would include both a written examination and an oral interview. This entire process takes a minimum of seven years to complete. It would only be after several more years' experience within a practice that a landscape architect would be competent to design all aspects of a million pound civil engineering landscape scheme.

The landscape practitioner should be able to take ideas, whims, impressions, feelings and notions and convert them into reality in 'bricks and plants'. This is very exciting, for it provides the enormous reward of personal creativity and moreover immortalizes vague dreams in practical, tangible reality. The landscape designer is, however, working with someone else's land and someone else's money. A duty of care is owed to the client to provide a level of expertise beyond that of the layperson and to exercise this expertise in a competent, precise and thorough manner.

Designers are different from both artists and scientists in the way they think and approach problem solving. This is because design requires 'convergent' thought. Most academic work is based on divergent thought – pulling things apart, analysing and categorizing information into its component parts. Convergent thinkers or designers use the information gained through the process of analysis but then put different elements together into a coherent and hopefully attractive and useful design solution. Artistic inspiration is the seed of this process but not the main factor as with fine art.

There are many elements to consider in landscape design, such as function, materials, aesthetics and context. Functions such as the provision of shelter, privacy, security and even delight must be provided for. Materials used to achieve these aims might be planting, paving bricks and slabs, stone walls, iron railings, water and so on. Aesthetics is about the principles of composition and beauty while context is a term used to describe the circumstances of the site.

The landscape designer

The subject of landscape design is a massive one, requiring knowledge about many related subjects. While it is clear that the job involves the preparation of an attractive design layout of someone else's land and the cost implications to them, it also requires the social skills to deal with both the client and the contractor, as well as statutory authorities and technical experts. Landscape design involves detailed data collection and assessment and pays regard to the utility, durability and aesthetics of the proposed landscape. Furthermore, much attention must be given to the overall co-ordination of the design elements to create a unified whole; a whole that must fit in with, and enhance, the wider landscape setting. Unlike many other design disciplines, the landscape designer must consider the fourth dimension too: the change or evolution of the scheme through time. Time is an important consideration to a new living landscape which is likely to require both management and maintenance to reach a satisfactory maturity.

Designing effectively for people's many and diverse needs demands a wide palette of knowledge and a landscape designer may find it beneficial to have some knowledge of horticulture, civil engineering, history, architecture, botany, geography, geology, soil science, meteorology, aesthetics, graphics, psychology and sociology, and a working knowledge of the uses and construction techniques for materials such as stone, water, wood, concrete, brick and metal.

All landscape practitioners have a professional responsibility to act in the best interest of their client, that is to say they have a duty of care to act as the client's adviser based on a thorough assessment of the client's requirements. However, it is only Landscape Institute members who are bound by the rather more specific professional code of conduct of that body. If the designer is commissioned to provide an administrative role during a contract between client and landscape contractor, then there also exists a duty for the consultant to act fairly between these two parties.

Furthermore, it is vital that the designer gleans all the relevant information about the site itself, the client's needs and indeed the client's budget. This will be achieved both by direct liaison with the client and by a full site examination and assessment. Client liaison will be successful only where it is possible to build a relationship of trust and goodwill. A mutual understanding of each party's obligations is essential and therefore so too is the agreement of the terms of the commission. Such terms relate both to the remuneration method, amount and when such a sum will become due. The terms will also define the scope of the services, the degree of authority to act on behalf of the client and the designer's liability.

Inevitably with a subject so vast there are specialist areas requiring specialist training, including the disciplines of both landscape managers (who possess a greater expertise in specifying maintenance and management practices and works) and scientists (who specialize in analysing and providing data on soils, plant and animal communities and other ecological and environmental factors). Above all, good landscape designers know how to source the specialist information that they require quickly, accurately and efficiently. Getting on with other people and having an effective chain of contacts is a great asset to a landscape designer whose work touches so many disciplines.

The service provided

The landscape designer's services are very similar to those of an architect but instead of being primarily concerned with the four walls of a building and inwards, they mostly affect all areas external to buildings (apart from interior planting). As such the landscape design services are primarily concerned with the design, construction and management of a site, involving a change in the quality and/or usage of it.

The works proposed may be either 'hard' landscape elements (such as walls, fences, paving surfaces and street furniture) or 'soft' landscape elements (comprising trees, shrubs, herbaceous plants, bulbs and grass). The different stages of the service provided are summarized briefly below and in more detail in Chapters 5–13.

Summary of stages of the designer's work

The various stages of the designer's work can be summarized as follows:

1. Approach to or from client.
2. Fee quotation and hopefully agreement.
3. Client liaison, formation of brief. Site appraisal.
4. Survey, production of base plan and evaluation.
5. Sketch ideas, notes and concept formulation.
6. Sketch scheme, client liaison and amendment.
7. Final design drawing, presentation standard.
8. Commence working drawings.
9. Prepare cost estimate.
10. Prepare bills/schedules of quantities. Choose contract.
11. Assemble tender documents and tender to an approved list of contractors.
12. Receive tenders and report to client.
13. Commence contract administration.
14. Certify final completion and client hand-over.

Earning a living in a tough world

Because you are reading this book, it is likely that you will be motivated enough to fulfil your ultimate goal, which might be to get a job and earn a living in the landscape industry as a designer, contractor or nursery manager, to improve your design skills or simply to be a more competent landscape practitioner. But the question is, what makes a competent landscape practitioner?

Imagination, sensitivity to clients and users of the site and some technical knowledge are, of course, essential, combined with a thorough knowledge of the subject. Such knowledge is learned during training and then practical experience. Spatial awareness is an essential capability in order to ensure that spaces of appropriate scale and character can be created. Good design concepts and ideas are necessary too. However, it would be a mistake to think that you cannot be a competent designer if you find yourself bereft of ideas at the beginning of a project, with a large sheet of white, blank paper set out before you. This phenomenon, sometimes called 'white paper shock syndrome', affects everyone, but some just hide it well. If you sit in front of the paper for long enough you will inevitably start to think about the scheme and 'doodle', and from first doodles mighty schemes can unfold. Often the first ideas are the best and are returned to even after looking at alternatives.

A good practitioner will also show great practicality and pragmatism in dealing with complex and often controversial situations. An ability to know when observations are preferable to judgements and to understand many viewpoints, to act fairly between parties and to avoid a 'high horse', entrenched attitude is vital. The client's criteria come first. It is the client's land, money and dreams that you are working with. However, at the same time, sufficient strength of character is needed to be able to offer confident advice, caution or encouragement to a client.

All of the above attributes contribute to competent landscape design but there is one singular characteristic that stands out above all else and that is simply *precision*. Precision in devising and communicating an intended design solution is arguably the most important discipline that the landscape designer must accomplish: the ability to define and describe precisely using a variety of media (drawings, models, schedules, notes and specifications) a complex design comprising many elements and then to specify precisely how to build each element satisfactorily. Such precisely defined information is necessary both to 'sell' to the client the designer's vision and to instruct the contractor to ensure a satisfactory construction of the design ideas, down to every last detail.

No single definition can be satisfactory where the subject is so diverse and complex as landscape design, but if there had to be such a definition to encompass the role and duties of the professional landscape designer, then it would be as follows:

> Someone who has the ability to define and communicate the precise quality and quantity of both workmanship and materials required to achieve an attractive and workable design solution that fulfils the client's needs and brief.

Whilst everyone makes mistakes, precision must be a primary goal of the professional practitioner if a reputation for competence is to be established. Of course, the achievement of the best possible design solution is our central purpose, but without precision a potentially good design solution is worthless. A reputation for competence is essential for success in the design world. Achieving it takes great concentration, commitment and a perfectionist approach to your work. It is made easier with both knowledge and experience but most of all by the personal drive spawned from a wish to realize your creative ideas in a way that you can be proud of, with the understanding that there is a creative idea at the root of even the most utilitarian design problems and solutions.

Pragmatic impartiality is also a mandatory attribute of the successful designer and cannot be separated from precision. We all have our individual prejudices and attitudes but it is the design professional's duty to supply accurate, impartial and independent advice to clients, to provide the full and precise facts, the full and precise range of options and to define fully and accurately the possible consequences. Then the designer must accept the client's decision.

Precision and clarity in written as well as oral communication are essential and the landscape designer must be able to provide for the client precise and conventionally set out contracts and specifications. Drawings and words must accurately specify and communicate the quality and quantity of workmanship and materials required to execute the design in a way that will ensure that a contractor can recognize and understand the information and build the scheme successfully. Errors in the quantity of both materials and workmanship may cause additional, unexpected expense for the client beyond the available budget, which may mean that the site cannot be completed as intended or on time. To earn a living as a professional landscape designer means attending to every detail and taking one's responsibilities very seriously!

CHAPTER 2

Design philosophy and aesthetic principles

WHEN approaching the design of anything, be it landscape, building, product, fabric, or whatever, it is important to think first about the *essence* of what you are designing, not just the function, the purpose, or the appearance but the real essence.

The essence of the subject

Illustrated below is what is meant by the 'essence' of the subject, not the obvious qualities but the very heart of what gives the subject its essential character.

Look at an orange

Facts

1. It is round.
2. It is orange.
3. It has a skin.
4. It contains pips.

Why is an orange so orange-like?

Not just because of the above qualities but because it is juicy, succulent perhaps!

Look at a landscape

Facts

1. Urban.
2. High proportion of paving.
3. Buildings surround a space.
4. Formal, avenue of trees, etc.

Why is a landscape so serene, grand or exciting?

Not because of the above qualities but more because of its cathedral-like trees.

Another example would be to examine the essence of a chair. Instead of an object with four legs, a flat seat and a back support, the chair should be seen simply as 'something comfortable to sit on'. From such a starting point could be spawned the design of a beanbag or indeed of a throne.

To define the essence the landscape designer must first determine the client requirements, the purpose and the functions demanded of the site. It is also important to determine the context, to define people's emotional responses to and feelings about the subject. Landscape design also requires an awareness and knowledge of the dynamics of scale and space required. By looking at the subject in this way, by abstracting it and conceptualizing it, the designer is able to break away from the preconception of the subject, work out the problems and find fresh solutions. The further removed from the preconception of the subject, the closer you can get to its essence, and the more you will be able to use your own creative ideas and create something unique and new.

There is nothing wrong, of course, with existing design solutions. Sometimes a design solution for a similar space and context to that which you now face may be entirely appropriate and can be used to good effect. However, existing designs can sometimes be prescriptive and predictable. Whilst they can be useful and sometimes the best, they will never be the most creative and expressive of your individual style.

Whatever the source of inspiration, thinking about the essence of a subject will help you make the right design decisions, appropriate to the subject's context, to its setting and to the client and site users.

Context/mood, *genius loci*

The meaning of mood

There are myriads of moods associated with a vast spectrum of contexts and landscape situations. Some sites are uncomplicated (a small urban back garden for example) and have relatively few moods associated with them. These simpler sites can well illustrate the meaning of the term 'mood' in design terms. An urban garden might simply require a mood of quiet contemplation, an oasis in the chaotic city in which to escape. The context for such a garden would still be 'home' and the need for privacy, security, shelter and rest would therefore apply. Some landscape contexts will have many moods associated with them, though these are rarely all in the same spot and therefore demand a compartmentalization of the site.

Typical moods inspired by the landscape include liveliness, contemplation and serenity, wonder and awe, majesty, suspense, mystery, delight, anticipation, astonishment, curiosity, intimacy of scale, grandeur of scale, formality, wilderness, power and prestige (the king of the castle), humility and so on.

Existing mood

If you have ever visited a high-rise housing estate in Tower Hamlets, London, for example, one that has received no external works improvements, you may sense an austere, harsh and even hostile mood about the place. This cannot be blamed on the residents: sometimes there is more community spirit surviving in such blocks than the landscape might suggest. It is the spartan drabness and the exposed, litter swirling, inhospitable microclimate of what are often seas of tarmac and grass that surround the blocks that are to blame. Add that to the often ugly and dwarfing façades of the blocks themselves and it is no wonder the place appears hostile. Use of the right landscape elements can bring about profound changes to this mood and many external works improvement schemes have transformed such areas (see Figure 2.1), while many more have yet to receive any of the meagre sums available for such urban improvement.

A visit to Sutton Place, near Guildford, Surrey, demonstrates how many different positive moods, from awe and wonder to simple delight (in the Paradise Garden), can be created in one site.

FIGURE 2.1

CONTEXT/MOOD – SARUM ESTATE, BURDETT ROAD, POPLAR, LONDON. Urban landscapes can seem harsh and hostile (top) but use of the right landscape elements (below) can achieve dramatic changes that enhance the sense of community spirit and *genius loci*, and reduce the feeling of austerity.

Proposed or required mood

Every landscape that is designed should provide for a desirable, positive mood, whether the simple delight of a sunny flower border or the deliberate sense of 'safe' danger of the hermit's grotto or theme park rides. Without positive mood a place would be without soul. There are too many soulless places in the modern world and it is the responsibility of landscape designers to ensure that there are no more. Every element of the design, from the shapes and surfacings to the enclosing structures, should contribute to and enhance the desired mood. A play area should have playful paving patterns, playful railings, playful street furniture as well as the usual arrangement of proprietary play equipment. A site's intended use will also infer a mood that may contrast or be in harmony with the character of both the site and/or the wider landscape setting.

The *genius loci*

The phrase *genius loci* means 'the spirit of the place'. Every place has a character, which may include the wider landscape or may be local to a street or village. The site itself may have its own character too, which may correspond to the wider landscape character or may indeed contrast with it. Some places seem to have more than a definable character; they have a presence, an atmosphere and sometimes seem to cast a spell over you. Such places may fill you with awe, humble you, inspire you and even scare you. Even sceptical readers will agree that some places are more striking than others (see Figure 2.2).

FIGURE 2.2 *GENIUS LOCI* – COXTIE HOUSE, BRENTWOOD. Every place has a *genius loci* which may relate to both the place itself and its wider landscape. This may range from an indefinable atmosphere to more definable and tangible characteristics: it is for the designer to interpret, adapt or create this spirit of place so that it makes the right kind of impact.

The question every designer wants to be able to answer is, how do you create places of presence that people will notice and will remember in the future, hopefully with affection? Ensuring that your designs express the appropriate mood is certainly part of the answer and will have hopefully therefore created for them their own *genius loci*.

Function

Function is the term given to the intended usage of a place and the client's requirement for a site, and several functions may need to be catered for. A picnic area, for example, would require seating and picnic benches and their provision would be an example of a physical function. Perhaps planting may be needed to frame a view or to screen an eyesore; these are examples of visual functions.

Types of function

There are three principal and very different types of function: activity-based functions, utilities and site problems.

Activity (or inactivity)
Activity-based functions include ball games, sunbathing, fishing, sitting, entertaining and eating outdoors, walking and riding (bicycles and horses), children's play, gardening, etc.

Utility functions

Utility functions include access points and routes, refuse provision, compost heaps, tool and other storage sheds, garages and workshops, services, pet housing and so on.

Site problem functions

The need to screen eyesores, provide shelter, ensure privacy and build in security measures are all site problem functions. Indeed such functions might also include the need to frame views in or out of the site (called isovists), mitigate pollution (including noise), provide deep shade, dappled shade or sun traps and so on.

Mood is the emotional response to a place, how one feels about it, and mood is associated with the site use and so is mainly concerned with the activity-type functions. Because it is the moods associated with these functions that are the key to our perception of the actual quality of a place, they demand great attention from the landscape designer. It is the mood of the place that makes you notice it, the function itself is noticed only when it does not work. Analysis of the appropriate mood is the key to creating good design ideas, in making the correct design decisions and in choosing appropriate materials. The style of graphic presentation to your client may also be varied to best express a mood and so sell your scheme to the client.

Mood without function

If you designed a space with a panoramic view over beautiful countryside intended as an attractive spot to stop, sit, picnic and take in the view, but there was no seating or picnic benches and the ground was undrained and waterlogged, it may still be a delightful place but it would not function as intended.

Function without mood

There are unfortunately all too many examples of functional design without mood. Much high-rise housing built in the 1960s and 1970s would fit this category but such examples exist in every period and style of architecture and the same can be said for landscape design, product design and so on. The Hulme Estate in Manchester once received a design award but has now been demolished. This brave new world of architecture was 'all mod cons' at the time, with provision of good sanitation, heating, hot and cold water and services, all of which the former 'slum' housing lacked. The one vital factor that the slums possessed and which was missing from the new high-rise housing was soul, humanity or mood. Providing for 'mood' humanizes functional design solutions. Much high-rise housing that characterized the 1960s and 1970s lacked colour, softness, ornament, shelter, privacy, security, private garden space, whimsical features and the gradual adaptations of buildings and their surroundings – of the kind that occur cumulatively through history.

The concept of context

Context is a concept, a picture or metaphor embracing many needs, moods, impressions, materials and aesthetic principles that collectively form an understanding of what a place is about, its character and its role in the users' lives. For example, a residential area, housing estate or a single dwelling would all share the same context, in this case encapsulated by the word 'home'. Privacy, security, community, intimacy of scale, friendliness, comfort, relaxation, play and so on are all positive words we perhaps would associate with the home, and which form the concept of the 'home' context.

For an urban town square, the context could hardly be more different. The context could be labelled 'public square', and busy ('buzzing'), lively, fun, sociable, grand scale and colourful would all be words that could be associated with the square. We might expect to see such a place occasionally be the scene of local events, and punctuated with monuments or memorials, all in an urban setting. These words and notions describe both functions and moods and are therefore 'key' words. They are often hard to define or to measure but nevertheless mean a great deal to all of us and have some potency in our everyday speech.

Determining the context of the subject or place is the first step to establishing a brief for the site and to defining the functions and moods, both of which must correspond to the context. If you are restoring a landscape, then it is the existing context that must be determined. This existing context will then be reinforced by analysing the functions, moods, materials and aesthetic factors that create the context. For new landscapes it is the desired context that you must define, and then determine the factors that will contribute to it.

Complex contexts

Some sites may have complicated contexts, that is to say there is sometimes more than one context required within one site, such as a public park. A variety of function and mood, materials and aesthetic principles will be needed to reflect these different contexts. Sometimes they will be opposites and necessarily juxtaposed due to limitation of space. For example, 'lively' sports areas may have to be placed next to 'contemplative' relaxation spaces. The contrast in both mood, scale and function will help to emphasize the difference between each one and as long as the functional requirements of the one type do not interfere with those of the other, then an interesting, varied and enjoyable park will be created for the needs of a wide range of people.

Materials (of construction) – hard and soft

Many factors affect choice of materials and the correct choice can make all the difference between successful translation of design ideas and complete failure of a scheme. If you build a wall using ordinary 'facing' house bricks, then the wall will soon crumble in the frosty weather and possibly fall down. Another example might be a client's desire for a 'secret' walled garden. If budget constraints forced a choice of a close board timber fence and gate, the right look would not be achieved. However, many materials are in fact substitutable without such loss of function and mood.

Materials and mood

The identification of the appropriate mood is often key to the decision-making process and materials help to convey the correct mood. The correct choice can enhance or detract from the desired aesthetic effect. This does not mean that mood determines materials in isolation, because adequate function is vital and so too is a correct specification, for the sake of durability. It is very difficult to retain an ideal aesthetic when paying proper heed to the disciplines of function and cost of materials and labour, but it is by no means impossible (see Figure 2.3).

Suitability of materials

It may seem an obvious point to make that the materials must perform or function as intended, but the case of a garden designer who replaced a manhole cover with a piece of chipboard and covered it with gravel to make it look more

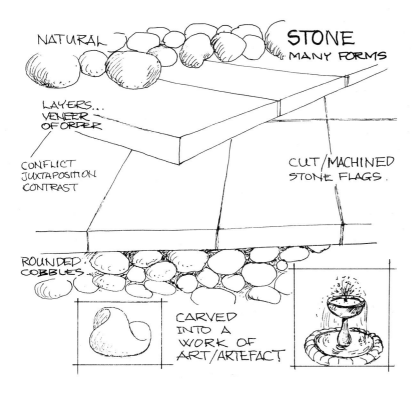

FIGURE 2.3 STONE. Stone is a perfect example of a material that can be used in a multitude of forms depending on the context or mood that is intended or existing. It can be used in its natural form as large boulders or rounded cobbles; it can be cut into precise ordered shapes and polished into crisp cubes and rectangles; it can be engraved or carved into works of art or used in architectural features and artefacts like fountains and sundials. The material itself can provide a unifying theme with gradual or sudden transition between its many forms.

attractive illustrates the importance of function. The board soon rotted and a man was injured when he fell through it down the hole. Material must be chosen for its suitability for the intended job, in terms of its size, strength and durability.

The material must be of a suitable grade and size, even after the appropriate type of material has been chosen. Another case illustrates this point. A pergola over a garden pathway sported cross beams that were designed marginally lower than the height of the client, who promptly injured his head when he tried to walk under them. Bricks may be a suitable material for pavements, steps and walls but there are bricks and bricks. Many are too absorbent of water to be fit for the intended purpose and are meant only for use in house building where they are protected by the roof eaves and gutters. Outside, exposed to the full fury of the elements, the bricks soon break apart, particularly after freezing weather.

Cost of materials

Cost is an ever-present constraint and there is no use designing a scheme using elements that demand materials that are too expensive for the available budget. The design could be carried out in separate stages and such phasing of the works can assist the designer if this option is possible but many elements are substitutable without necessarily compromising the quality of the design. The need for enclosing elements in a garden along a boundary, for example, may suggest a wall but in some circumstances a suitably attractive and durable alternative would be to use a close board timber fence with a trellis fixed in front of it and attractive climbers grown over it. This option would be significantly cheaper, but often still satisfactory aesthetically and functionally. It is a useful exercise to think of some alternative enclosing elements that perform different functions. Some enclosing materials may prevent physical access across a boundary, but allow views in and out of the space, for example railings or trellis work, whilst a brick wall presents both a physical and a visual barrier. Clearly, a brick wall is a far more expensive barrier than a close board fence and yet both provide a physical and a visual barrier. The same comparison can be made of paving materials with York stone flags being more expensive than brick paving, which itself is more expensive than concrete slabs, but even these are dearer than natural materials such as hoggin or pea shingle. A balance must be struck between cost and durability.

Aesthetics

Aesthetics essentially means the study of how to make something more visually pleasing. Aesthetics is a complex concept that encompasses the broad principles of composition and arrangement, and there are many aspects to this highly controversial subject, from the philosophical and metaphysical, to the more mechanical laws and notions that govern the use of line, plane, proportion, focal point and so on. The controversy is caused by the polarization of two very different viewpoints. Some think that 'beauty' is in the eye of the beholder – entirely subjective and purely a matter of individual taste – while others believe that there exist fundamental principles and laws governing the creation of beauty, which are definable, measurable and therefore objective. It this dilemma that has caused much philosophical debate over the centuries.

The truth must lie somewhere in the middle. Robert Pirsig, in his book *Zen and the Art of Motorcycle Maintenance* (published by Vintage, 1989) in his analysis of logic itself, found that beauty, and indeed the concept of quality generally, are subconscious discoveries or 'events' and that analysis of them as being subjective or objective takes place only after the event has happened. Concepts like beauty and quality are phenomena somehow encountered on life's leading edge, as our senses (and perhaps our souls) literally cut through time, like the stylus point on a record groove, amplifying raw creativity. Only later do we stop to ponder the concepts and try to define what is indefinable. Feebly we attempt to describe some magical event, saddled with cumbersome words in the here and now, when the event happened before here was here and now was now! Perhaps this event occurs in the subconscious mind before we become conscious of the here and now. This thought nicely introduces a theory propounded by Carl Jung who was a pioneer in his theories of the unconscious mind. He believed that part of our unconscious mind is collective and common to all humankind. If this is true, then it would help to explain why we so often share the experience of beauty, and why so many long-standing aesthetic and compositional laws and principles have been developed over many centuries.

There is one level where beauty must be collectively shared and therefore some aesthetic principles (even if of a relative nature rather than strictly objective) can be defined and used with confidence. There is also a level where the perception of beauty is far more superficial. At this level, beauty is more a matter of individual taste and is influenced by cultural factors, including socio-economic grouping, fashion, personal philosophy and outlook, peer pressure, nationality, media and so on.

Aesthetic principles

The notion that the perception of beauty is collectively shared is of value to the designer and it is embodied within many of the principles of 'good' composition, taught in fine art but equally applicable to landscape design, and indeed to other design subjects. These principles include the laws of perspective, focal point, balance, framing of views, the concept of background, middle distance and distance, relative complexity, novelty and so on. Aesthetics then is a subject of study that encompasses the basic elements of landscape composition, central to the world of painting. There are detailed principles for the use of line, plane, point, shape, proportion, scale, pattern, colour, texture, tone, etc. Aesthetics further embraces some complex concepts important to the practising of landscape design, and such theories include the creation of space, defensible space, prospect-refuge theory, and edge effect. Some of these concepts have been hypothesized precisely because they define collectively shared responses to the spaces and places around us. Such collective responses may well stem from the different ages of man's psychological and evolutionary origins (forest dweller, hunter/gatherer and settler) and indeed possibly from the psyche of man's destiny: the desire for betterment (adventurer, industrialist and technocrat).

Style

The style of a landscape usually refers to the nationality of the landscape or to a period in history. We therefore refer to Japanese style, or English Romantic, Baroque, Italian, Moorish style and so on. Style may also refer to the degree of formality in the design, ranging from the severely formal grandeur of Versailles to the soft, sweeping and rounded landforms of the English Country Park in the Capability Brown style. On a more intimate scale, style may range from a symmetrical courtyard with tidy box hedges and bedding plants at one end of the scale to the romantic chaos of the cottage garden at the other.

Style is therefore a more superficial aesthetic phenomenon than the main aesthetic theories listed above and is a product of cultural influences. It is therefore a subject relating more to personal or national taste than to a more collective or universal aesthetic.

Theories based on man's origins

Carl Jung believed that there were elements of the human subconscious mind that were collective to all humanity, and the deeper into the subconscious, the more universal were its contents. Any landscape designer who entertains this concept in any way and intends it to influence design decisions must first be able to identify those elements of our perception of the landscape that are collectively shared.

The works of the late Sir Geoffrey Jellicoe (including his famous book *The Landscapes of Man*) came close to fulfilling this requirement and the information below on the first three ages of man is based on his work. But this book offers both new interpretations of his analysis and a different view on the later stages of man's evolution. An alternative vision of the evolutionary ladder is presented below.

Man in the forest

'Man', 'he' and 'his' are intended here to include womankind. At the earliest identifiable stage in man's evolution, he lived in the forest. His home was made in glades in the trees and he was entirely dependent upon the forest for his food and shelter. This 'age' of man's evolution produced a deeply collective, subconscious response to the landscape, inherent in us all (like a race memory) even

today. It gives us a sense of awe and humility when standing under a grand avenue, similar to the experience felt in a cathedral, where the stone columns equate to the tree trunks and the stone vaulting to the tree's arching branches overhead. The arched shape of the stained glass windows at the end of the cathedral is like the arch of light at the end of the tunnel of trees (see Figure 2.4).

We get a sense of comfort, calm and security when sitting under tall foliage, epitomized by the Thomas Hardy poem 'Childhood Among the Ferns'. In my view this stage in man's evolution is closely linked to emotions that relate to the home and of spiritual recreation. Such feelings might include security, cosiness, calm, peace, privacy, awe, humility and spiritual well-being.

Man the hunter/gatherer

Later in man's evolution he left or cut down the forest and organized into social groups to lead a hunter/gatherer existence in the open plain. The landscape was characterized by wide open spaces which could be surveyed from hilltop fortresses, advantageous both for hunting game and for spotting enemy tribes. The wide open spaces give us a liking of grand views and the hilltop fortresses make high ground particularly desirable if the spot is both concealing and sheltered. Such landscapes are epitomized by the work of the famous landscape gardener Capability Brown, with his sweeping park-like landscape style (see Figure 2.5).

FIGURE 2.5 MAN IN THE SAVANNAH. We collectively find panoramic views attractive – especially where animated with grazing animals, little copses and trees that harbour wild game. Such landscapes were (arguably) characterized by Capability Brown with wide prospects, clumps of trees, grazing animals and fish-filled lakes.

The universally desirable, secluded hilltop picnic spot is featured in many eighteenth-century English landscape paintings, perhaps with a soldier standing resolutely gazing out to the horizon and wondering, no doubt, what the future battle has in store for him. Such paintings might typically contain a pretty lady in a white floaty dress sitting gracefully on the picnic rug pouring tea, with the couple set nicely against some sheltering background trees, conveniently planted to frame the view and create a perfect composition.

The very real desire to seek out places that are concealed and yet have a wide panoramic view forms the basis for J. Appleton's 'prospect-refuge' theory, where he argues that man will find places enjoyable where there is a good prospect and yet safe refuge from potential hazards. He hypothesizes that this enjoyment stems from a subconscious race memory from our hunter/gatherer past, when tribes built fortresses on hilltops such as Chanctonbury Ring on the West Sussex South Downs, where they could look out in relative safety from fierce animals and hostile war parties. Indeed the children's game 'hide and seek' is arguably based upon this phenomenon. The emotions associated with this shared landscape response include security or refuge, a sense of power or control, of grandeur, expectation or anticipation, wonder at the panorama and also perhaps reflection too.

Man the settler

Man soon learned to domesticate animals and grow the best plants for food, clothing and ornament. The birth and development of agriculture led to the building of towns and the polarization between town and country areas. Organization and order were the hallmarks of this age, and our desires to grow houseplants, vegetables, garden flowers, tidy lawns and keep pets all stem from it. From the tidy row of carrots to the controlled chaos of the cottage garden, the age of the settler has given us a taste for order and design in the plant and animal kingdom.

Man the adventurer and the destiny of mankind

Sir Geoffrey Jellicoe described other ages of man including the age of the voyager, which is supposed to have given rise to man's adventuring spirit. However, the alternative view that this spirit of adventure was there from the start, and that it was the catalyst for man's development in the first place, makes this the deepest and most collective emotional response and therefore of profound importance to the designer.

Rather than just another phase in man's development, the adventuring spirit of man is the key to his destiny. It is the spark that caused him to make journeys into the unknown, to explore, to invent, to create and to conquer. It is the very spark of creativity itself. The thrill of adrenaline, challenge and dare are its powerful manifestations. So too is the desire to seek danger, even the safe danger of the theme park ride, or to flirt with the supernatural, even if in the harmless form of a Hammer horror 'B' movie or a garden grotto. The industrial revolution, the electronic revolution and all invention is part of this great drive for betterment and more excitement and indeed so is the great striving for improvement in the worlds of politics, creative arts and, most sublime of all, philosophical and spiritual enlightenment.

Whether the landscape designer sets out simply to create mystery in landscape designs or to introduce more specific intellectual challenges to the client/users, awareness of this collective subconscious phenomenon will help to ensure that final designs are more interesting and even more exciting, amusing and fun (see Figure 2.6).

FIGURE 2.6 MAN THE VOYAGER. There is something in all of us that strives for betterment and adventure, and in life, work and play mankind is always pushing back the frontiers. So there is room for landscapes that challenge the mind and one's spirit of adventure.

Space and edge effect

The creation of space

The creation of space is a fundamental principle of landscape design. This principle is most simply described by imagining the site to be a solid jungle to which you take an axe and make clearings for each of the required activities, the size of the clearing being suited to the activity. The mass of vegetation then encloses and defines the cleared voids or spaces (see Figure 2.7).

FIGURE 2.7   SPACE CREATION, COXTIE HOUSE, BRENTWOOD. Space is a key element of landscape design and the right effects cannot be achieved without enclosing features such as plants, or hard features such as fences and walls. Enclosing outdoor space is comparable to enclosing indoor space through subdivision into rooms.

Enclosure of space

Space implies enclosing elements, which can be either hard or soft materials. These enclosing materials define space in width, depth and height. Examples of enclosing elements include walls and fences, trees and shrubs, kerbs and steps or combinations of them all. Some are more successful than others, enclosing physically but to a limited extent visually or vice versa. The use of materials to enclose space can be likened to subdividing an area into rooms, such as those of a house, each with a different size, character and use, to accommodate the client's intended functions and activities. This is not to say that such spaces cannot be unified as a whole design, in the same way that the diverse rooms of a house can work together in a unified way to form a cohesive and comfortable house. Clearly, though, the designer must be aware that a space for quiet contemplation will be different in size and character from one for rowdy recreation-

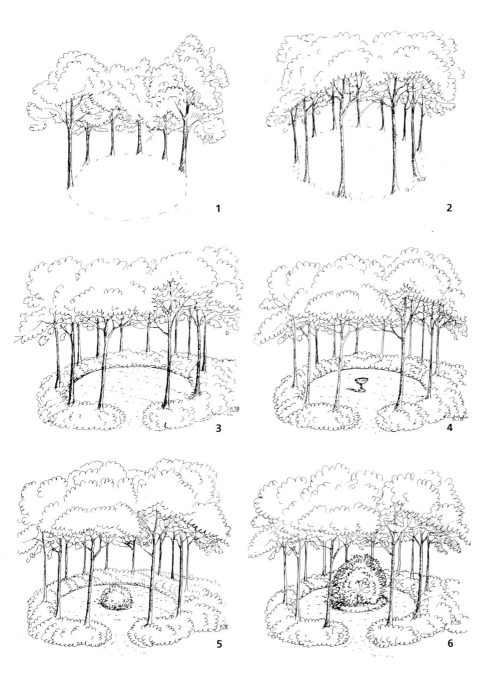

FIGURE 2.8 SIX STAGES OF SPACE CREATION

1. A piece of flat land can be encircled by points to imply an enclosure – here with trees which also define height as well as breadth and depth.

2. A piece of land has now been completely encircled with trees to imply a void within.

3. The implied space is now reinforced and further defined by shrub planting. An entrance point is not enough to destroy the sense of space.

4. The insertion of a small artefact in the centre of the space serves to enhance the creation of space and focus the eye on the centre of the space.

5. A larger object is inserted into the centre of the space (in this case a bush) and this has become intrusive on the space and has changed it.

6. The shrub has grown and has now destroyed the space in its original form, instead creating a corridor space around the edge of the central feature.
It is clear that the scale of any artefact/object within a space is crucial to the success of space creation.

al activities like ball games, as in the same way your lounge will not have the same character and size as your bedroom, kitchen or dining room.

Size of the space
The size of the space must be large enough in width, breadth and height to carry out the desired activities comfortably.

Utility of the space
A space must be sheltered enough, private enough, secure enough and so on to carry out the desired activities in comfort and without adversely affecting other adjacent activities.

Mood and style of the space
The mood and style of the space can be determined by manipulating its shape, contents and materials, and by the functions and utilities required. The space must convey the right atmosphere, for example playful, contemplative, serene, lively, awe inspiring, mysterious, novel, strange, intriguing, delightful, sombre and so on or perhaps a mix of compatible moods. A sequence of spaces might reflect all these moods in turn.

Breaking up the space
Any feature placed in the middle of the space will have a dramatic effect on it. The size of the feature is crucial to this effect. If the feature is small, it will become a focal point within the space. Any larger feature will begin to look out of proportion and make the space look mean and peripheral to the object. Larger still and the feature will actually break up the space and create a new corridor-type space around the edges (see Figure 2.8). Planting beds are the most common cause of this type of space destruction in gardens, often placed centrally. If an item is placed in the centre of a space to create a focus, then to succeed this object must be of a proportionate scale to the size of space. A small garden space might be improved by a central sundial or small sculpture but would be destroyed by a large monument, tall shrub bed or raised terrace.

Degrees of enclosure
If a hedge were placed around a small lawn it would define a space, but it would be far more successful if there was some low planting in front of the hedge to soften the junction between the vertical plane of the hedge and the horizontal plane of the lawn. Indeed the more layers to the enclosing planting, the more definite and successful will be the feeling of space. Several layers (at least three) are best for defining spaces and these layers are like those found naturally occurring in native woodland ecosystems. Designers mimic these natural layers by using ground cover plants at the front, a medium-height band of shrubs to tall shrubs and trees at the back, the tree canopy eventually defining the height of the space.

Some types of enclosure material provide effective shelter within the space while others actually cause wind eddies there. Wind permeability is the key factor here, hedges and plants providing better shelter than walls because they slow down the wind as it passes through. Walls cause the wind to jump over the wall and create swirling eddies on the lee side. Snow fencing uses this principle, by allowing some wind through the fence, slowing down the drifting snow and depositing it before it reaches the roads.

Edge effect

Edge effect is the phrase used to describe the universally shared feeling of unease at being in the centre of a large space and of comparative well-being when you have your back literally against a wall (or similar structure). The advantage in defensive terms of your back being against the wall, so that you

FIGURE 2.9 EDGE EFFECT. The person in the middle of the cricket ground will be feeling more exposed and vulnerable than the person at the edge of the pitch, who will be feeling more exposed and vulnerable than the person sitting on the front row of the stands. The people who are most at ease are the couple sitting at the back of the stand with their backs against the rear wall concealed in the shadow of the roof. People will always feel more at ease at the edge of spaces, particularly large open spaces.

can see all potential enemies in front of you, is, of course, the origin of the well-known phrase. If you were to be placed naked in the middle of a cricket pitch (not by choice as a streaker), you might feel a little vulnerable and self-conscious. You would almost certainly feel better if you were fully clothed sitting on a chair behind a table at the edge of the green. This phenomenon of preferring places where you are close to an edge of a space is closely related to J. Appleton's 'prospect-refuge' theory and it is easy to understand why one feels more at ease in smaller, 'human scale' spaces, where you can be reasonably close to an edge at all times (see Figure 2.9).

There can be no doubt that we collectively and subconsciously like edges. They make us feel more secure and are our touchstone and reference point. We feel safe on a pavement because of the kerb, though this might be only 100 mm high. Of course, this love of edges applies mostly to the edges of relatively flat spaces, especially large ones. A severe drop like the top of a cliff edge or canal edge may well have the opposite effect.

We judge the size and scale of space according to a subconscious notion of the human scale. Some spaces we may describe as grand or even dwarfing, while others will seem more intimate in scale and feel secluded and cosy, but these may, of course, conversely appear cramped or poky. Emotions that we would associate with being in the centre of a large space such as a sports green or Moscow's Red Square might include conspicuousness, vulnerability, exposure and insecurity. Emotions that we might associate with small spaces, closer to a 'human scale' or when near to a definable edge might include cosiness, shelter, privacy, security and sanctuary.

Compositional design principles

Once the broad-brush principles have been sorted out, it is time then to concentrate on the more immediate design factors that affect the design composition itself. Such principles concern the identification and manipulation of line, plane, change of level, focal point and so on. Many of these principles arise from the laws of composition in fine art and yet in the design of three-dimensional landscapes an awareness of these compositional laws is crucial to producing a successful result.

The fundamental difference between a painting and a landscape is that the landscape is viewed from ever-changing viewpoints, is more interactive and has to look good from all sides in all seasons.

Balance

Balance is the term given to a state of visual comfort that helps to ensure that a composition is attractive. This comfort is derived from viewing objects belonging to a scene that collectively have the same or similar visual dominance or gravity on both sides of the picture, that is to say right and left of the vanishing point (see Figure 2.10). The vanishing point is the imaginary point at which any parallel lines in the scene will join as they go away from the viewer into the distance. Therefore if the objects collectively on either side of this vanishing point have widely different visual gravity, then the scene is said to be in imbalance and will look uncomfortable.

When using symmetry, a perfect balance is achieved, but this can on occasions conversely have a dramatically uncomfortable effect. (See also 'Symmetry', p. 36.)

Unity

Unity is the term given to a design or composition when the many component parts all relate to each other harmoniously to create one whole. Objects can easily appear to have been scattered randomly in a design (or indeed in a landscape) and are therefore disharmonious; each object seems to be free-floating and have little connection or sense of belonging to other objects

FULCRUM
POINT OF BALANCE.

BALANCE
ACHIEVING BALANCE IS ESSENTIAL FOR GOOD COMPOSITION: PERFECT BALANCE OF SYMMETRY CAN SOMETIMES CAUSE **DUALITY** WHERE THE TWO HALVES PULL APART DISHARMONIOUSLY.

FIGURE 2.10 BALANCE. The landscape is attractive when the eye is able to find a balance point, which should coincide with a focal point.

or to the boundaries (see Figure 2.11). The goal of achieving unity does not require all the elements of a landscape to be similar or co-ordinating, even where a series of spaces contains completely different themes. Unity can instead be conveyed in the way the shapes fit together, the sequence of the spaces or by the structural space-defining elements.

The proximity of one vertical object with another can be crucial to achieving unity – the objects aid perspective and balance and enhance the composition, providing a vital tension between them.

It is essential that any spaces be related to the boundaries of the site so they appear to belong there. Relating shapes, spaces and objects to the boundaries can be achieved in the first stages of a sketch design by perceiving the site as a solid block in which you carve out cells. These cells are the spaces in which the proposed objects and activities can take place. These spaces can be given a structure that relates each one to each other like a honeycomb. Unlike a honeycomb, each space must be different to maintain interest, and changing the shape or size or content of the spaces (or all three) will secure the interest of the users throughout their journey through the site.

FIGURE 2.11
DISUNITY/UNITY. The spaces and objects that make up a design must be in unity with themselves and with the boundaries and setting or they will appear to 'float' disharmoniously, apparently unconnected and disunited or unrelated to the site.

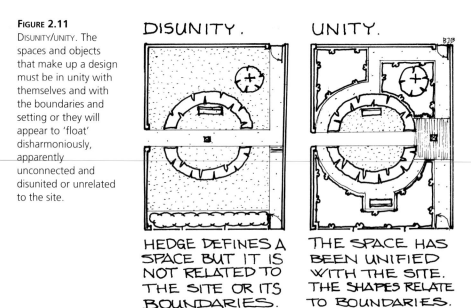

DISUNITY.   UNITY.

HEDGE DEFINES A SPACE BUT IT IS NOT RELATED TO THE SITE OR ITS BOUNDARIES.

THE SPACE HAS BEEN UNIFIED WITH THE SITE. THE SHAPES RELATE TO BOUNDARIES.

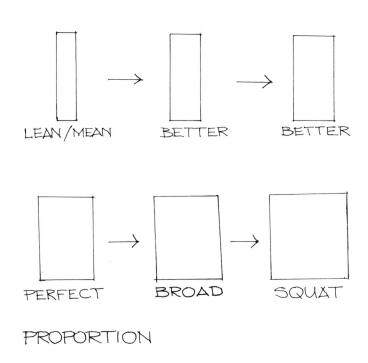

LEAN/MEAN   BETTER   BETTER

PERFECT   BROAD   SQUAT

PROPORTION

FIGURE 2.12 PROPORTION. If ever there were proof that beauty does not just lie in the eye of the beholder alone but is somehow a more tangible thing, then proportion must be one of the best pieces of evidence. Perfect proportion lies diametrically poised between the lean/mean and the squat.

Proportion

The relationship of width, height and depth is crucial to the success of any objects in the landscape. If a window is too wide or too tall it will appear 'out of proportion' and disharmonious. Any landscape object, whether a paving slab or a swimming pool, must have great attention paid to its proportions in order to achieve a successful design. To ask what is the correct proportion is to invite centuries of debate, but it is always easier to point out things that are out of proportion, as disharmony is more noticeable than harmony (see Figure 2.12). The Golden Section is a fundamental proportional formula invented by Leonardo da Vinci and used in classical architecture. It is a guide to proportions that we collectively find harmonious because they relate to the human form. We develop a subliminal sense of the 'normal' and we seem to have an in-built sense of what is and what is not in proportion, perhaps developed from when we first familiarize ourselves with the human form and with the world at large, and so come to judge an object in terms of how much it differs from this normal human scale or human proportion. The Golden Section is a rule of thumb that helps to satisfy our needs for proportion to please the eye.

Line

Line is essential to a composition to lead the eye into the scene from foreground to distance. Lines can convey the mood of a place – they can be flowing and sensual or articulated and geometric; they can be strong, intermittent or implied. They can help to unify a design by linking different elements and they can relate to the grain of the landscape and to its contours (see Figure 2.13). Line of vision is important in traffic safety as well as aesthetically in creating focal points, vistas and panoramas.

Plane

The horizontal plane is essential for the existence of space and the vertical plane for defining space. Changes in level can also help to define space. It can be interesting to imagine that the surfaces and surfacing materials used in a land-scape are like layers of floating lily pads, just as continental plates float on the earth's crust in stratified layers. Different surfaces can be designed to appear to be sliding out from under others to mimic the earth's tectonic plates. Such ploys encourage an awareness of thickness in the user, as well as width, height and depth, so adding interest. Changes in level, punctuated by vertical elements, can create powerful visual effects.

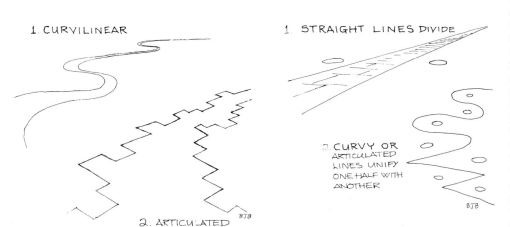

FIGURE 2.13
CURVILINEAR/ARTICULATED AND STRAIGHT LINES. Straight lines cut across a composition and create a division, though they can be used to create powerful perspective in formal design to highlight a focal point. Curvilinear or articulated lines have the effect of uniting the areas on each side of them, the former in an organic and natural way, the latter in a geometric, manmade, abstracted way.

FIGURE 2.14 FOCAL POINT, COXTIE HOUSE, BRENTWOOD. Points in a landscape act as punctuation and can orientate the spectator, sometimes directing the eye towards a particular focal point in the distance such as the gate and landscape beyond in this photo. Points can imitate natural features such as the avenue of trees imitated in this picture by the pillars of the pergola.

Point

The incidental points in the landscape act as punctuation or landmarks, helping us to orientate ourselves (see Figure 2.14). They act as full stops to rest the eye at the ends of vistas and we refer to them then as focal points. Points can be positioned to lead the eye into the scene, aiding the effect of perspective. Artists take great care to position points in the landscape with crucial distances between each other in order to create vital tension, lead the eye into the picture and engineer a good composition.

The vertical point is more powerful visually than the horizontal, just as an exclamation mark is a more powerful figure of literary punctuation than the full stop.

Feature

When a point in the landscape is chosen because of an intrinsic beauty or intellectual interest in itself, rather than purely as a visual mark to rest the eye, then that point can be described as a feature (see Figure 2.15). As with any other

FIGURE 2.15 FEATURES, COXTIE HOUSE, BRENTWOOD. Points in a landscape such as a doorway or sculpture attract the eye and allow the eye to rest. At first the feature attracts attention away from the general landscape context which surrounds it but when it has become familiar it is simply a focal point that attracts the eye for a short while before it moves on to the next arresting feature.

point in the landscape, the eye may rest on the object, but instead of travelling onwards, a feature will hold the interest for quite some time. Awareness of the landscape as a whole ceases during the time the feature is being examined under close scrutiny. Such objects may be sculptures of great beauty or cleverness, puzzles, artefacts of antiquity, symbolic stones, supernatural or religious icons, or even scientific instruments. The landscape can enhance the interest generated by a feature by providing a setting to which the feature can belong. Eventually, as the novelty factor gives way to familiarity, the feature will be viewed as an ordinary focal point, the eye roving onwards until the next feature is encountered.

Accent

Emphasis can be given to a part of the site by the use of accents, whether large mass plants or one small group of plants or contrasting hard materials, especially surfacings. Accents serve to enhance views or create bright spots in a scheme. Contrast is the essential tool to achieve a successful accent and such contrast must be sharp, which demands boldness. For example, bright masses of flower

FIGURE 2.16 ACCENT. The use of an architectural sword-leaved plant amongst ground cover material creates accents which draw the eye into the view. The large existing tree also provides an accent in this landscape.

ACCENT
SHRUBS CAN BE USED TO CREATE
ACCENTS - "PUNCTUATING" THE LANDSCAPE.

BJB

or foliage in otherwise sombre tones, a vertical streak in a horizontal theme, a round form in a place of straight lines (see Figure 2.16).

Contrast

Contrast is the tool that allows us to gauge the relative difference between things. Contrast enables us to build up a picture of the average (or the normal) and this standard can then be compared to the unusual. We might say that a texture is coarse because it is so compared to the normal, or perhaps it is fine compared to the normal. Powerful contrast between elements creates a visual tension that grabs the eye. Yet the extremes have by definition to be compared with our subliminal notion of the normal to be effective and therefore the juxtaposition of two extremes will visually cancel each other out and ruin the effect.

Greatest contrast, and therefore greatest impact, is created by juxtaposing an extreme with the normal. For example, coarse texture next to medium texture is more powerful than coarse next to fine. The two poles can be brought together to good effect, but only when the 'normal' is there to act as the touchstone. Contrast can be used in design in two ways. Either by starting with one extreme and gradually changing it until it becomes normal, then perhaps going on gradually to become the other extreme, with imperceptible transitional stages, or by using the cataclysmic break, with one extreme dramatically juxtaposed with the normal (see Figure 2.17).

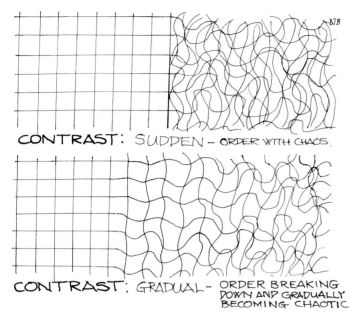

CONTRAST: SUDDEN - ORDER WITH CHAOS.

CONTRAST: GRADUAL - ORDER BREAKING
DOWN AND GRADUALLY
BECOMING CHAOTIC

FIGURE 2.17
CONTRAST. Sudden contrast can be very dramatic unless the forms or textures are both extremes, when they can cancel each other out and blunt the effect. Sometimes a more powerful effect can be created by placing two extremes either side of an intermediate texture or form.

Contrast is the food of life for landscape designers and it is used from the most abstract stages of design conception (perhaps contrasting the geometric with the natural) through to the detailed design of practical components of the landscape (such as colour, form and texture, whether plants or paving stones).

Rhythm

The repetition of objects, surfaces, colours or materials in the landscape provides a rhythm that brings a sense of unity and continuity. This rhythm is arguably as important to the success of a landscape composition as audible rhythm is to a piece of music (see Figure 2.18).

Sequence

The repetition of similar (or related) but not identical objects, surfaces, colours or materials in the landscape can provide a useful co-ordinating theme for a succession of compartmental garden spaces. The spaces themselves could be designed to

RHYTHM

FIGURE 2.18   RHYTHM. The repetition of the same theme, feature or pattern can create a feeling of ordered rhythm within a landscape composition.

FIGURE 2.19   SEQUENCE AND EVENT. Points, features or artefacts can be placed in the landscape so that there is a sequential change along a route or line of sight; where they are the same then instead of sequence you can create rhyme. Event means the arrival at an interim or final feature, which might be anything from a terracotta pot to a sundial or the façade of a building.

FIGURE 2.20
PANORAMA. Few people would disagree that it is pleasurable to view a wide open landscape, preferably from a high advantageous viewpoint, where the horizon imperceptibly bleeds into the sky. It can sometimes be better to remove vegetation rather than plant it, though panoramas are often best framed with trees to focus the view.

change shape and/or size in sequence, for example in a garden, the intimate and ordered spaces of the patio and formal lawns gradually changing to the wild open spaces of the copse and meadow beyond (see Figure 2.19).

Event

In any sequence of spaces, there will need to be an incident or event that provides a sense of arrival. Generally, the more grand the spaces, the more grand the event must be, though Sir Geoffrey Jellicoe (inspired by the surrealist painter Magritte) designed an anticlimax at the end of a grand parade of Grecian urns in the grounds of Sutton Place, near Guildford in Surrey. The event can be a purely visual display, such as a dazzling array of flowers, or an intellectual puzzle, perhaps a maze. Sculpture can provide either a visual, intellectual or even spiritual event. Events have taken many forms through the ages of garden design, from grottoes, temples and other follies to figurines and fountains (see Figure 2.19).

Vista, view and panorama

Of course, views occur naturally on the ridges of high ground and such wide open views we call panoramas. Narrower, directed views are called vistas, and landscape design through the ages has paid much attention to the creation of such vistas and views. By framing the natural scenery the views are enhanced and are brought closer to the ideal of the perfect landscape composition, according to the rules of composition of landscape painting. The landscape designer must focus the viewer's eye on the best parts of the prospect and block out the rest using a frame of screening materials, be they trees or buildings. Even panoramic views are best framed with trees, and indeed many eighteenth-century landscape painters used trees both as a device to frame a beautiful view and as a dark backcloth to set behind the figures (see Figure 2.20).

Backcloth

The use of a dark tone of uniform and fine texture, such as a yew hedge, against which to set bright flowers or ornamental shrub borders is a traditional tool of the landscape designer. The backcloth of dark tone and fine texture provides a perfect contrast to coarse-textured, bright and silvery foliage and flowers (see Figure 2.21). Yew hedges are therefore the traditional backcloth to the herbaceous border.

Where a formal hedge is not appropriate, tall-growing, evergreen shrubs can be used for structure planting, not only to ensure year-round definition of space but also to provide a backcloth to the ornamental species in front.

Backcloth is also a term used in landscape planning, referring to the wider landscape, to the habitat type and local flora species associated with it,

FIGURE 2.21   BACKCLOTH, COXTIE HOUSE, BRENTWOOD. Backcloth provides contrast and increases the definition of a space, setting off the foreground. Here *Sambucus nigra Aurea* (golden elder) are contrastingly set against a dark group of trees including a Copper Beech.

which gives a distinctive character to the local and regional landscape. Some sites will be viewed against this local landscape backcloth and all materials will have to be chosen with great care to blend in with it.

Texture

When light strikes a surface, every hollow, protrusion, undulation and pimple casts a shadow and the rougher the surface the bigger the shadows. The bigger the shadows, the greater is the contrast between light and dark and so the coarser the texture. As mentioned earlier, generally we develop a sense of the 'normal' through our experience of the world and we judge objects we encounter in comparison to this subliminal concept of the normal. We therefore describe a surface as coarse or fine textured when compared to normal or medium texture.

With plants, the bigger the leaves, the bigger the shadows and so the coarser the texture. Any extreme of texture, be it coarse or fine, is visually more powerful than medium-textured objects because it is an extreme. Therefore the designer should only put coarse textures next to fine textures if they are both juxtaposed to the normal or medium texture as a reference point. This is because the two visually powerful textures (coarse and fine) will cancel each other out unless a medium texture is present to act as a foil.

It is a fact well known to artists that coarse textures advance towards the viewer, visually, and fine textures recede away from the viewer. The landscape designer can use this knowledge to manipulate the effect of perspective by using texture gradients. This effect can be further enhanced by use of colour.

Colour

Colour is a vast subject but the basic principles for the landscape designer are as follows. Yellows and reds are warm colours and they advance towards the viewer visually. Blues and some greens are cold colours and recede from the viewer. This is another way to manipulate the effect of perspective. Spaces can be made to appear longer or shorter by using a colour gradient. This effect is enhanced when combined with the use of a texture gradient.

The nature of colour is too complex a subject to go into in greater detail here but colour must be used with great care. All colours have opposites that are called complementary colours. Taking the simplest example, using the primary colours of blue, red and yellow, for yellow the complementary colour would be the sum of the other two colours – purple. Therefore green is the complementary colour for red, and orange the complementary colour for blue. Complementary colours can contrast strikingly with their corresponding colour, particularly in the presence of an intermediary, but can equally cancel out the intended effects of their corresponding colour if mixed too finely for the scale of the site. They are best used boldly.

The most important aspect of colour for the landscape designer is not, however, the use of flower colour (important though these are as accents for short periods) but concerns the contrast of leaf and stem colour. Foliage and stems are ever present and mistakes in their use are consequently far more serious. The main colours encountered are silver, green, purple and gold. These colours are deliberately listed in order. The two colours in the centre (green and purple) are both foils, associating well with both gold and silver as well as each other. The colours at both ends, silver and gold, are very powerful visually but they must usually be used singly with either of the 'foil' colours adjacent to them. Gold and silver look horrible together without a foil colour between them.

Scale

Scale is the term used to describe the size of a space or site (see Figure 2.22). Large scale in landscape terms is associated with grandeur, conjuring up images of Versailles and the natural wide open country typical of some English counties, like Hampshire and Wiltshire, and of the British coastline or the National Parks with their often mountainous terrain. These powerful natural backdrops can make a complex design appear over-designed or even twee. The same design in another setting might, however, be entirely successful.

Small scale in landscape terms refers to more intimate spaces, like courtyards or external spaces or rooms extending out from a dwelling house. Far more intricate designs can be very successful for such small-scale sites. Japanese gardens create great detail in very small spaces and miniaturize everything to create an almost model-world aesthetic. A rockery is perhaps a more familiar parallel. The rockery is designed to appear as a miniature cliff face or rock outcrop. The illusion relies on the use of small alpine plants and dwarf conifers for its success. When these grow too big, they destroy the illusion and the stones look insignificant and artificial.

Deliberate confusion of scale can be a powerful and/or even humorous

FIGURE 2.22
INTIMATE/GRAND SCALE. Small spaces enclosed with busy planting and with detailed static paving designs and interesting artefacts as focal points create comfortable and cosy outside rooms that are pleasant to dwell in and pass through. Bold, simple designs with clearly defined boundaries, lines, points and planes with spectacular features and events can create powerful landscapes which can be humbling and demonstrate great power over nature and the landscape.

INTIMATE SCALE   GRAND SCALE.

device, well used by surrealist painters and anarchic comedians. For example, an office desk shown on a wide panoramic beach, with a 'city gent' sitting behind it (*Monty Python's Flying Circus*), and a door pictured opening on to a cliff top (Magritte).

Pattern

Pattern is about the arrangement of shapes, the shape of spaces and the juxtaposition of surfaces and materials. The design process cannot escape involvement with pattern, whatever the site or brief. Landscape design has similarities to fashion design in this respect and landscape designers often talk about the fabric of the landscape, the patchwork of fields. The relationship of one shape to another is crucial to harmonious design, both in the way the shapes interact and in the distance between the shapes. Sculptors like Ben Nicholson were inspired by the apparent tension created by the proximity of one geometric shape to another.

Patterns can be symmetrical or asymmetrical, formal or natural, flowing or geometric and so on. Pattern alone is too superficial a device in landscape design to ensure a successful design but a design that has no thought given to pattern will appear confused and weak.

Boldness of pattern used should be determined by the scale. Just as a big-patterned wallpaper in a small room will make the room appear even smaller, a bold pattern in the landscape is best used for large-scale situations, where fussy patterns would be lost. Fussy, fine-textured patterns in small-scale spaces make these spaces appear larger and the detail of the pattern can be appreciated, seen as it is from a close proximity. Bold patterns in small spaces look crude and make the space feel cramped.

Symmetry

Symmetry is a tool of the formal or classical garden design. It has by definition a centre line in which one side is mirrored by the other (see Figure 2.23).

Symmetry lends itself to use with geometric shapes but organic shapes can also be arranged symmetrically to good effect. Symmetry is a formal way to achieve a balanced composition, as long as the shapes are well proportioned.

There are many examples of symmetrical buildings, including most of classical architecture but less fortunately including the 1950s semi-detached house. This latter design illustrates the problem of 'duality' which is associated with some symmetrical designs. The two halves of a symmetrical design visually pull away from each other. This does not normally matter if the shapes are well proportioned but if the

FIGURE 2.23
SYMMETRY. Symmetry is the use of the mirror image and is employed in formal landscape design. Care must be taken to avoid duality, which occurs if the proportions are wrong.

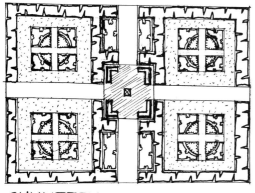

SYMMETRY.

shapes are out of proportion, perhaps just a little broader than the 'perfect proportion' (as defined by the Golden Section), then the pulling apart is exaggerated into a battle, which is very uncomfortable for the viewer. The Golden Section is the best guide for avoiding duality.

Asymmetry

Asymmetry was a tool much prized by eighteenth-century landscape painters and the Romantic Movement in garden design, where compositional balance was created by using different-shaped objects of similar visual 'gravity' rather than the mirror image of symmetry. Asymmetry creates an informal natural aesthetic. Flowing lines are used to lead the eye into the composition and the result is by definition more mysterious and less ordered or tame (see Figure 2.24). Very powerful visual effects can be created by juxtaposing symmetry with asymmetry, either bleeding one into the other with an imperceptible join or contrasting one against the other dramatically. The deliberate use of classical symmetry breaking down under assault from the wild forces of nature has been a source of inspiration for painters throughout history.

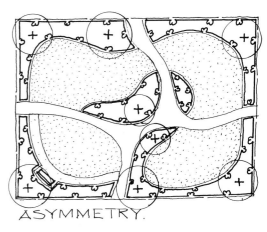

ASYMMETRY.

FIGURE 2.24
ASYMMETRY. Asymmetry employs the irregular arrangement of objects, lines and spaces and these can be designed in geometric patterns or organic forms.

Geometry, shape and their interaction

Formal

Formal landscape spaces are often designed largely on the basis of three main shapes: the square, the circle and the triangle. These shapes can be stretched and dissected to form arcs, arches, rectangles, diamonds, stars, ovals and so on. Squares and rectangles can be fitted with arches at each end and these can be smaller than the width of the square or rectangle to produce a step. Shapes can be fitted within shapes, as with knot gardens and Italian sunken gardens. The shapes of the spaces can be mirrored in the vertical elements too, in the walls, hedges and trees, creating clipped topiary walls or more intricate effects and features.

Inspiration for how to interact these formal geometric shapes can come from many sources: from architecture, fine art, graphic design and even industrial and textile design. One designer once used a motif on a baby's bath mat for the basis of a formal garden design!

Informal

Informal shapes can be employed to mimic natural form, and inspiration can come from the study of biology and natural history. Biological shapes are never weak or aimless. They are mostly definite, strong and often curvaceous but sometimes geometrical too, especially if the circle is employed. Whether your inspiration comes from fine art, the shape of an animal, amoeba or the path of a meandering river, the informal line is a powerful aesthetic. In their most pure forms the circle or the oval may be used. These shapes seem to be at home in both formal and informal aesthetic camps. The circle is particularly powerful, being both symbolic of our biological origins (the ovum) and perhaps also symbolic of our destiny; the perfect geometric form, holding a pure and powerful symmetry. The use of the curve in design provides a softening effect on architectural form, whether you choose to use the regular oscillations of a radio wave or the wild convolutions of a writhing snake as your pattern book. The intersection of such lines is difficult to design successfully, though for both aesthetic and practical reasons it is best achieved at 90 degrees. Narrow points can look harsh and toothy, difficult to protect and are difficult (and sometimes impossible) to plant or to turf, especially where concrete foundations extend beyond the edging treatments on either side.

Mixed formal and informal

One of the most exciting design situations is the linking of formal and informal aesthetics together. This mix can be achieved in a gradual way, one form gradually transforming into the other through a wide spectrum of traditional stages. Or it can be done in a 'time lapse', staccato way, with a gradual change up to a point, then a sudden but small change followed by another gradual change and so on until the opposite form is reached. A dramatic contrast of these equally powerful but opposing aesthetics can also be used. This contrast is hard to achieve without one style looking out of place, or indeed one aesthetic cancelling out the effect of the other, but, if successful, this method can present one of the most powerful of images. To make this strong contrast work, there has to be a purpose, a concept, a *raison d'être* for their juxtaposition. The 'earthquake' approach can be used with a zigzag break to contrast one against the other like two land masses colliding on a fault line. The 'layer' approach is another alternative and can show the different aesthetics peeling away at different layers as if the patterns had been eroded away to reveal different strata below.

One of the most enjoyable aspects of the design profession is to play around with geometry, shapes and their interaction, solving functional and spatial problems in the process. There are still many unexplored questions about the underlying psychological reasons why some shapes work better than others and why some shapes can appear in proportion, while others only slightly larger or smaller look squat or mean in comparison. How much this is to do with the notion of human proportion and scale as explained by the Golden Section and how much is of cultural origin will, perhaps fortunately, remain in hot debate for decades to come.

What is good design?

'Good design' is an often used phrase but it is hard to get two people to agree exactly what it means. It must at least require as a precondition that the landscape designer has correctly identified the functions, activities and the moods required by the client and also those pertaining to the site context. Furthermore, 'good design' must include the choice of suitable materials for the construction and must achieve the right aesthetic criteria. The preconditions of 'good design' could therefore be summed up as follows, though it must be remembered that although these factors must be present, possessing these factors alone is not enough to ensure a design is 'good':

1. Functions and moods
    The landscape designer must solve functional problems and provide for desired utilities and activities. Moods that correspond to the desired functions must be created.
2. Specifying materials
    The designer must balance the conflicts of quality, cost, durability and appearance in choosing the materials of construction and must also ensure that they are put together in a suitable arrangement.
3. Aesthetics
    The attractiveness of the design will depend on the skill of the designer in using aesthetic principles and ideas to create a pleasing arrangement.

CHAPTER 3

Hard landscape design

Context and mood

HARD landscape simply means brick or slab paving, timber fencing, metal railings, structures and any other non-living external landscape element.

Once the initial design layout has been completed and approved by the client, the designer can then begin to work out the detailed choice and arrangement of fencing types, walling, paving pattern, edging and street furniture. Some of the early detailed hard landscape design may be carried out before working drawings are commenced as it is often important to know, for example, the fencing types, wall design and the paving patterns required, even at the sketch design stage. Just like the initial design stage, the starting point for choosing and detailing hard landscape elements must take account of the context and mood of both the general area and of the immediate site.

The character of a place is defined by the relative proportion of hard elements, such as dwellings and other buildings, to soft elements such as farm land. Recreational space and street furniture are all indicative of these character definitions. It can be useful to pin down such contexts when choosing materials in order to avoid specifying inappropriate ones which then look incongruous on site. The temptation exists to overuse standard details for convenience without sufficient thought to the setting, especially with smaller elements such as knee rails, bollards and cycle barriers. Metal detailing may look very smart in an urban area but appear brutal in a rural and rustic setting. Listed below are some of the indicator factors that define the more measurable landscape contexts, so that the right decisions can be made on choice of hard landscape detailing. The character and choice of various hard elements and street furniture will vary widely between the following contextual categories. For example, a railing chosen for a country house or market town might well be the curved top continuous bar railing often found around paddocks in such areas; in the inner city, railings might vary between traditional vertical bar railings with ornate finials to modern architectural metalwork in its various forms.

Urban
- High proportion of hard space
- High density of buildings
- High quantity of shops, financial centres, public buildings and offices
- High quantity of street furniture
- High proportion of public space and low quantity of private outside space
- Low quantity of agricultural activity

Rural
- Low proportion of hard space
- Low density of buildings
- Low numbers of shops, offices, financial centres, public buildings
- Low quantity of street furniture
- Low proportion of public space and high quantity of private land
- High quantity of agricultural activity

Suburban
- Medium proportion of hard space
- Medium density of buildings; a few shops, but low numbers of public/financial buildings and offices
- Medium amount of street furniture
- High quantity of private land but some public space
- Low to medium agricultural activity; some paddocks

This simplification of the subject can act as a rough guide. For a more subtle assessment of context and mood the individual landscape designer must rely on more subjective findings. Often, though, such assessments which precede hard landscaping decisions cannot be undertaken in isolation but are formed by a consensus of local planning officers, conservation officers, architects, local residents, councillors and members of professional advisory bodies such as English Heritage, the Countryside Commission and so on.

The following discussion represents a selection of some of the terms that can refer either to the existing site (and/or locality) or to the intended use of the site, and which help to determine or form a context or mood and to guide the choice of appropriate hard landscape detailing. It shows how hard landscape elements can influence the character of a site.

Modern/high tech

A modern or 'high tech' approach might conjure up images of smoke glass office blocks and tubular steel construction, suitable for business parks and financial centres such as the London Docklands. Bold, smart, paving patterns and colours are the order of the day, with modern architectural metalwork and street furniture designs.

Traditional/historic

We might describe a landscape as traditional or using historic reference where materials and patterns complement historic and traditional buildings. Such period architecture might be medieval, Jacobean, Georgian, Victorian, Edwardian and so on. Some urban features might be described as Dickensian and examples of these might include finial tipped railings, with York stone pavements or perhaps stock brick walls and paths, granite sett roads, with cart ruts and so on. Such materials are associated with images of quaint cobbled streets, with bow window fronted second-hand bookshops and an 'Olde Tea Shoppe' or two.

Static

FIGURE 3.1
DIRECTIONAL PAVING,
COXTIE HOUSE,
BRENTWOOD

Static is a term often used for spaces with no definite route or direction to them, such as urban courtyards – self-contained spaces where the paving does not dictate a path or indicate a flow of traffic. Paving patterns might include basket weave brick paving or circle patterns.

Directional

Directional is the term used where spaces are linear or else have one or more routes. Paving (or even fencing) patterns will be pointing in a direction to suggest flow of traffic (see Figure 3.1).

Fussy or complex

Fussy or complex patterns or detailing are generally associated with small-scale spaces, though not always so. Complex or fussy paving is very popular on the Continent, especially in Germany, and not only in the smaller private courtyards but also in the public squares. Generally busy, complex paving patterns, such as fan pattern paving, should be used for smaller squares or when travelling slowly (see Figure 3.2).

Simple or bold

Simple or bold patterns or themes tend to be most effective for large-

FIGURE 3.2   COMPLEX, PATTERNED PAVING, CAVELL STREET GARDENS, STEPNEY, LONDON. This scheme won a BALI award of merit in 1990.

FIGURE 3.3   SIMPLE, BOLDLY PATTERNED PAVING (1997 BALI award-winning site at the Arundel House Hotel, Cambridge).

scale spaces and where viewers are travelling at speed – walking or driving. The larger the space and the faster the pace, the simpler and bolder the pattern should be (see Figure 3.3).

The function of hard materials

The function of hard landscape materials is simply to be suitable for the intended site use. Therefore it is important to choose materials that are fit for the purpose but which also reflect the context of the site and help to create the required mood.

'Function' (when referring to hard materials) can be understood to mean character (i.e. aesthetic function), performance and durability. Any hard landscape material chosen for a design should be suitable in all these ways.

Character

A surface suitable for a road in a rural park might suggest a rustic material such as gravel for reasons of both mood and context and yet also suggest tarmac for reasons of performance and durability. The solution perhaps would be blacktop with gravel dressing, which might blend in well with the rural context and help establish a mood of easy paced, rustic charm.

Performance

Performance relates to actual suitability of the material for the job, the relative size or thickness of the material required and the method of construction.

Durability

Durability relates to the performance of the material over time and after weathering. For example, untreated timber for a fence will quickly rot. This is especially so where the post meets the ground. Bricks laid for a free-standing wall or for paving must be frost resistant or they will absorb water and break apart (sapping) in frosty weather.

Selection of materials

It is generally sound advice first to choose the appropriate material and only then determine the right specification or exact type of material. For example, you may choose a timber fence for a front garden, then decide on the timber thicknesses required to prevent damage from vandalism and finally choose a protective and decorative stain. This approach will help to ensure that the appropriate character of material is chosen before performance is considered, which will avoid starting with a compromised, second-best, belt and braces aesthetic.

Methods of space enclosure

Enclosing hard landscape elements may include walls, fences and gates, railings, bollards and structures such as pergolas. Because landscape design theory centres on the concept of space creation, as explained in Chapter 2, the hard landscape elements listed above are arguably the most important in landscape design. This is doubly so because vertical elements are far more powerful visually than horizontal ones.

Hard materials are the first-line method of space definition. They sometimes define space with a certain harshness that requires softening. In a room we apply decoration, ornaments and furniture to soften them and in the garden we use plants. Plants can be used to appear like hard materials, for example a close-clipped yew or box hedge (see next chapter).

Walls

Walls can be made from many materials, such as brick or stone (natural or reconstituted), and can be single or double skin. They can be laid dry or with mortar binding, coursed, uncoursed or semi-coursed (see Figures 3.4 and 3.5).

Corners are often treated with special care for structural strength and visual enhancement. With stone walls, the corners require larger stones and these corners are called quoins (see Figure 3.5). Some stones, called through stones, will go right through a wall. Brick walls may require piers for stability and have movement joints at approximately 12 m intervals. Half brick walls (105.5 mm wide) require bays and returns for stability, but being so thin they are the least expensive.

FIGURE 3.4 SEMI-COURSED WALL, ST EDMUND'S HOSPITAL, BURY ST EDMUNDS.

FIGURE 3.5 STONE WALLS

STONE WALLS

COURSED - EVEN

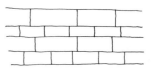

COURSED - VARIABLE

SEMI-COURSED ; DRY STONE

SEMI-COURSED; MORTAR JOINTED

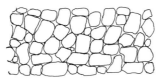

UNCOURSED - DRY STONE

UNCOURSED; MORTAR JOINTED.

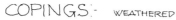

COPINGS:

WEATHERED

UNWEATHERED

THROUGH STONE

SECTION   ELEVATION

SECTION   ELEVATION

QUOINS.

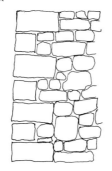

BATTERING.

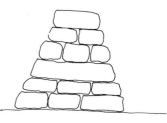

Snake walls also can be half a brick thick, their snaking course providing the stability, but these walls are more expensive because of the extra length required to provide the snaking effect. Half brick walls with piers are dearer still, almost as expensive as a solid one brick thick wall. One brick walls (215 mm wide) are much more stable (and suitable for walls up to 1.8 m). Such walls at 1.8 m need piers (or returns) and these should be built no further apart than 12 m, as should the movement joints, which can be 10 mm wide and filled with fibreboard and mastic sealed.

Walls over 1.8 m high require designing specially by a structural engineer using a formula which takes into account wind pressure at the height proposed, mortar strength, bonding, foundation depth, the angle of repose and so on. The wall is then built to meet the criteria calculated and may, for example, be constructed to a height of 1.2 m at a brick and a half wide (330 mm) and from 1.2 m up to 2 m at one brick thickness. Alternatively, the wall may be constructed with concrete blocks, which have a brick skin, the blocks perhaps strengthened with steel reinforcement bars. If brick is used for the whole wall, then the brick and a half thickness wall can be linked aesthetically to the one brick thick wall built on top by using special bricks with sloping ends or sides, known as plinth headers or stretchers, respectively.

There are essentially three ways to prevent damage to a wall from the worst of British weather. First is the correct choice of brick using a frost resis-

FIGURE 3.6   COPINGS

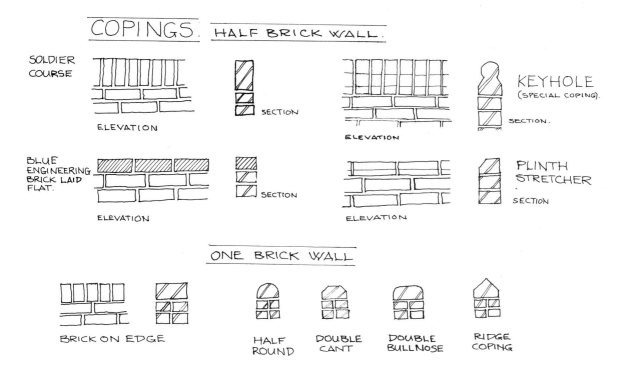

tant brick, such as the 'Purple Multi Stock' from the Ockley Brick Company. Second is the use of an adequate coping, which could be brick on edge or the use of special coping bricks. Such special bricks might include bull nosed copings, ridge, cant or half round coping bricks (see Figure 3.6). Third is the use of a damp proof course (DPC) to prevent rising damp damaging the bricks through the effects of freezing and thawing. The DPC is best achieved with three courses of Class A engineering grade bricks.

Bondings

Brick walls can be laid in a number of different styles, and these are known as bondings, each with a special name (see Figure 3.7).

Stretcher bond means that the bricks are laid with their stretcher face outermost and overlapping the bricks below. This is the strongest bond because the weight of bricks is spread evenly outwards. It is also the least interesting bond.

Flemish bond involves alternate 'stretcher' and 'header' bricks and can only be used for walls one brick thick (225 mm) or more. This is probably the second most common bond of brick.

English garden bond can similarly be used only for walls one brick thick (or more) and only where there are at least five courses of brickwork below the coping and preferably not less than seven (485 mm and 626 mm respectively). It involves two or more, usually three, rows of stretcher bond bricks followed by one complete row of headers.

FIGURE 3.7 BONDING

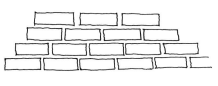

STRETCHER BOND

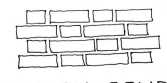

FLEMISH BOND

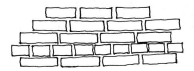

ENGLISH GARDEN WALL BOND.

Pointing and mortar

Both the bonding and the brick type can be enhanced (or indeed detracted from) by varying the pointing of the mortar (see Figure 3.8). Pointing is the term given to the trowel finish given by the bricklayer to the mortar joints.

Bucket handle, the most common finish, gained its name because it is often achieved by using the rounded handle of a bucket. This form is easy to achieve and therefore the cheapest, though not very exciting.

Flush pointing is less commonly seen and means that the mortar is flush with the surface of the bricks. This gives the least dominance possible to brick bonding.

Recessed pointing is the rarest but most striking type of joint. Some of the mortar joint is raked out to form a trough. This is the least satisfactory in terms of weathering, as water can lie in the joint and freezing can cause damage to the mortar and/or bricks. The deep shadow cast by the joint makes this a visually powerful method of pointing, which really emphasizes the bonding and colour of the bricks.

Figure 3.8 Brick wall pointing

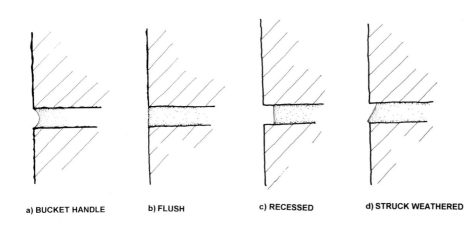

a) **BUCKET HANDLE**   b) **FLUSH**   c) **RECESSED**   d) **STRUCK WEATHERED**

A good compromise is the struck weathered joint in which the mortar is sloped back to shed water, giving a good shadow and emphasizing the quality of the brick, while keeping water out of the joint.

Mortar joints can be coloured, either to blend in with or to contrast with the bricks. A poor contrast is generally better than a poor match, and a good contrast is better than anything. Darker mortars again help to emphasize the bricks and bonding.

The addition of lime to the mortar reduces the strength of the mortar, which helps to preserve the softer bricks. The mortar in some old brick walls, such as many in the medieval village of Lavenham, Suffolk, is made from raw lime, which is appalling for the bricklayer's hands but provides those thick, pebbly white soft mortars of really old walls that are hard to match using modern mortars. As re-pointing is often required on old walls, specialist restoration firms are required to match such raw lime mortars. The mortar strength must be just right to avoid hastened weathering of the brick itself.

Timber and wire fences

Fences come in many forms, for many different purposes. Most fences will be made from timber, but some modern architectural metalwork could technically be defined as a fence rather than a railing. For convenience all metal fences are included under the 'Railings' section.

Rural fences

Rural fences are not solely used for livestock but also have such diverse purposes as protecting new plantations from rabbits or controlling public access in country parks. Such fences are commonly post and wire fences using pressure-treated rounded softwood timber posts (see Figure 3.9). Either rounded or barbed wire is then fixed to the posts in three equally spaced bands with galvanized 25 mm staples. When these three wires are fixed under great tension they are called straining wires and are tightened using devices called winders or tensioners

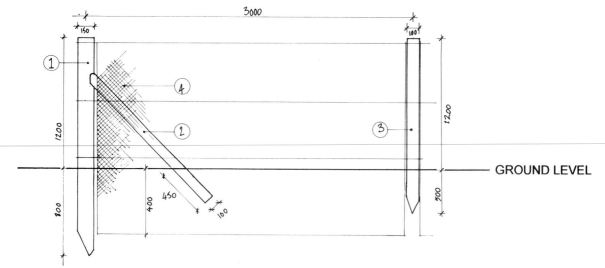

Notes

1. Timber straining posts 2 m length, min. 150 mm diameter.
2. Timber struts 100 mm × 100 mm × 1.6 m length.
3. Intermediate post 1.5 m length, min. diameter 100 mm.
4. Galvanized 'rabbit mesh' to be strained along fence, stapled to timber with 25 mm galvanized fencing staples very tightly spaced to ensure no gaps. Mesh to be let into the ground 400 mm.

Rabbit fencing/stock fencing
Stock fencing identical in design, substitute chicken mesh with squared mesh stock netting and do not let into ground.

General notes
• All dimensions in millimetres
• Do not scale from this drawing
• All timber to be CCA treated to BS 4072 Part 1

FIGURE 3.9 RABBIT OR 'STOCK' FENCING

incorporated adjacent to end posts, which being larger (at 150 mm diameter) are called straining posts. The straining posts are driven into the ground (but sometimes fixed into concrete) at 30 m intervals on straight runs, with supporting struts notched in and to intermediate spacing posts of 100 mm diameter.

There are various sizes of squared or round hole mesh, used for pigs, sheep, rabbits, chickens or for general forestry containment. Rolls of these mesh types can be fixed to the straining wires, using the 25 mm galvanized staples.

This fence, particularly with chicken or rabbit mesh, is ideal for protection of trees/shrubs from marauding wildlife. The mesh will need to be let into the ground to a depth of 400 mm to prevent rabbits digging under the mesh. The height of fencing will be determined by its usage, though most rural fences are 1 m to 1.2 m high, with the exception of deer fences which are over 2 m. Such fences present a relatively low cost way of producing a physical barrier but they do not create a visual barrier. They do have a psychological effect, however, and can be useful to deter access by the public on to private land or on to wildlife sanctuaries.

Ornamental fencing

Ornamental fences start with the visually semi-transparent palisade fencing, with its hit and miss spaced vertical bars of timber nailed (using galvanized nails) to parallel horizontal rails, themselves fixed to 125 mm square posts.

Picket fencing is virtually the same, but the bars are narrower (75 mm wide), have a pointed top and are more flimsy generally, using smaller timber sections, and are usually no more than 1 m high. This fencing is commonly used to encircle front gardens and may be stained with tinted preservative or painted white. Any preservative stain that is used should be checked to ensure that it is not phytotoxic, that is, not toxic to plants, particularly if it is intended that shrubs or climbers will be growing on or in front of the fence.

The close board timber fence is the most visually impermeable, comprising a series of thin broad boards (125 mm wide, 14 mm thick) nailed to three horizontal triangular arris rails which have been pointed to fit slots cut in the 125 mm square posts. This fence can be up to 2.5 m tall and as low as 600 mm, although the most usual height is 1.8 m (see Figure 3.10).

Timber posts are normally used, sawn softwood and pressure treated, although oak or even precast reinforced concrete posts are also sometimes used. It is important to have a good foundation, which may be a concrete block

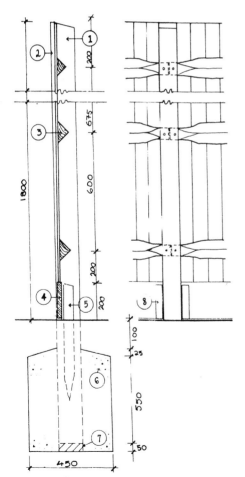

Notes

1. 125 × 125 mm sawn softwood post 2.7 m max. Centres morticed as required.
2. 90 × 14 mm tapered to 7 mm softwood boarding. 50 mm galvanized nail fixing.
3. 75 × 75 mm triangular arris rail shaped to fit mortices. All to be nailed through with 60 mm galvanized nails.
4. 200 × 25 mm gravel board in sawn softwood.
5. 50 × 50 × 600 mm stump let into ground at centre of gravel board.
6. 450 × 450 × 750 mm C:20:P mix concrete foundation.
7. 100 mm dia. pipe section to form stand for post of galvanized mild steel.
8. 50 × 38 mm bearers nailed to post and gravel board with galvanized nails.

General notes
- All dimensions in millimetres
- Do not scale from this drawing
- All timber to be CCA treated to BS 4072 Part 1

FIGURE 3.10 CLOSE BOARD FENCING

C:20:P mix 750 × 450 × 450 mm or a metal shoe. The best of all, and also the most expensive, is a metal shoe holding the post 30 mm above ground, with a spoke set well into a concrete block foundation, below ground level.

The area of timber most prone to rot, when posts are set into shoes below ground or into concrete blocks, is the point where the post enters the ground. At this point the post has exposure to a continuous combination of dampness and air, ideal for fungal growth. By keeping the post above ground in a metal shoe, the post is high and dry, and therefore this is certainly the most durable construction.

Weather protection will greatly increase a fence's life expectancy. Most fences can benefit from post caps being fitted to the top of the posts; the caps are larger than the top of the post and are roof shaped with a four times weathered top. A roof shaped capping can also be fitted over the ends of vertical boards and horizontal rails, adding further protection for palisade fences. Other forms of weather protection involve the treatment of the timber. Softwood timber can be treated by pressure exposure to tannin compounds, then can be stained in addition by hand, with two to three coats of wood stain preservative, such as Cuprinol or Sadolin 'Classic', to the preferred colour.

Gates can be made to match all the above fences but must possess very heavy duty galvanized latches because gates easily warp in changeable weather and only the most robust latches will be able to cope without seizing up. Galvanizing will protect the ironmongery against rust. The most commonly used gate latch is the Suffolk thumb latch, although hand-ring drop latches are also common. Drop bolts, too, are commonplace, in conjunction with gate latches, especially for vehicular double gates. Some gates can be constructed with self-closing springs or hinges. These will, however, cause the gate to bang, which can cause a nuisance to people living close by. Rubber or foam buffers can be fitted to minimize this problem.

A good visually semi-permeable fence can be made by using trellis panels, fixed to 125 mm square posts, with climbers, especially the evergreen Japanese honeysuckle or ivy, grown up them. Most of the above timber fences can be mounted on brick walls using posts bolted to the brickwork or to metal brackets bolted or set into the brickwork. Designing your own timber fence can be a rewarding exercise if you don't like the conventional ones.

Railings and metal fences

The most durable fencing is undoubtedly achieved from metalwork, galvanized mild steel being the most commonly used material, though stainless steel is occasionally used as are certain grades of aluminium and, of course, wrought iron.

Mild steel is the least expensive metal, although it must be galvanized to prevent rusting. Railings can be painted all sorts of different colours, although adhesion of paint to newly galvanized steel can cause problems. There are only two ways of achieving a lasting result:

FIGURE 3.11
BRUNSWICK ROAD,
LONDON, BEFORE AND
AFTER BRICK PLINTH AND
WAVE PATTERN RAILINGS

1. Use a 2 pack acid etch primer, undercoat and gloss × 2 or even 3 coats – obtainable to RAL or BS colours.
2. Use a special paint such as Sika Inertol 'Icositt' Highbuild 5530 that is obtainable to RAL colours.

Railings or metal fences can be of any height, but are mostly under four metres. The development in metalwork design over the last ten years by pioneers such

FIGURE 3.12 BEDE AND SARUM ESTATE, POPLAR, LONDON, BEFORE AND AFTER RAILINGS

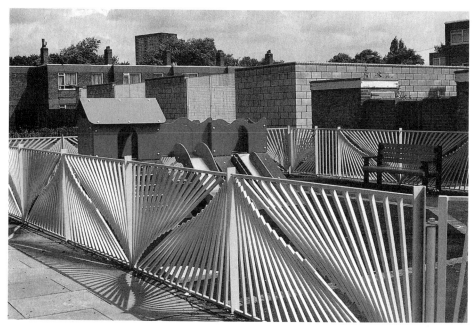

as Stuart Hill, formerly of Claydon Architectural Metalwork Ltd, has ensured that just about any conceivable pattern can be achieved – even visually impermeable railings. Such railings include wavy bar, leaning bar, musical and all manner of wacky and innovative designs (see Figures 3.11 and 3.12).

Railings can have smaller-sectioned posts for equivalent strength to timber and smaller foundations that can be simply a block of concrete in the ground. Some of the simpler railing designs are therefore less expensive than some timber fences. Generally, railings are more urban in character but not exclusively so.

Metalwork is much less susceptible to weathering than timber fencing and therefore gates can sport much finer-quality locks and latches and, indeed, lockable latches requiring keys can be fitted to make external areas as secure as internal ones. More often normal drop latches are used or else sliding bar fasteners. Both can be drilled to receive padlocks. Self-closing springs can be fitted or self-closing hinges used. Gate designs can match the railing design or be of ornate or unusual design as independent entrance features (see Figure 3.13). Vertical bar railings and gates often have the top of the bars decorated with metal finials, which come in many shapes and sizes, from round balls that screw on to spear head finials to fish tail designs and so on.

A common and inexpensive metal fence is the chain-link fence. Varying in height from 1 m to 4 m, this fence is a common surround to tennis courts

FIGURE 3.13 RAILINGS AND GATE, CAVELL STREET GARDENS, STEPNEY, LONDON

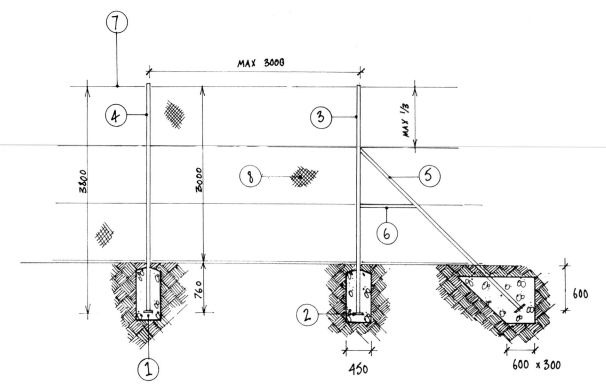

Notes

1. C:20:P mix concrete foundations.
2. Steel base plate 150 mm × 150 mm × 8 mm welded to struts, straining posts and intermediates.
3. Rolled steel angled straining post 60 mm × 60 mm × 8 mm × 3800 mm including proprietary steel base plate (2) bolted to struts and braces with 1 No. 5 mm diameter heavy duty galvanized nut and bolt per joint.
4. Rolled steel angled intermediate 50 mm × 50 mm × 6 mm × 3800 mm including proprietary steel base plate (2) set into concrete foundation (1) at maximum 3 m centres.
5. Rolled steel angled strut 45 mm × 45 mm × 5 mm × 3800 mm including proprietary steel base plate (2) set into concrete foundation (1).

6. Rolled steel angled brace 45 mm × 45 mm × 5 mm × 1000 mm bolted to (3) and (5) with 1 No. 5 mm diameter heavy duty galvanized nut and bolt per joint.
7. Line wire 3 mm diameter galvanized and plastic coated at minimum 1000 mm vertical centres.
8. Chain-link wire mesh to comply with BS 4102, galvanized and plastic coated to 50 mm.

General notes
- All dimensions in millimetres
- Do not scale from this drawing
- All metal parts to be galvanized and painted with Sika Inertol 'Icositt' Highbuild colour black to manufacturer's instructions

FIGURE 3.14   CHAIN-LINK/STEEL POST FENCE

and yards, compounds and car parks (see Figure 3.14). Usually a mesh of 100 mm square in galvanized mild steel wire is used but a more attractive finish can be achieved using plastic-coated chain-link, available in black, green or brown.

Bollards

Boundaries can be described simply by a line of bollards. These define an edge, such as a busy road, but allow free flow of pedestrian traffic and are, of course, visually transparent. Bollards are most often used to protect the pedestrian from

METAL AND TIMBER BOLLARDS.

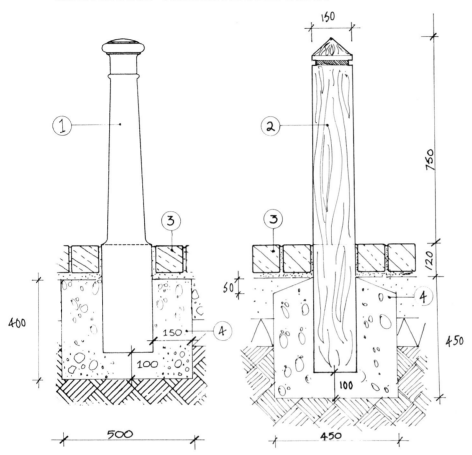

Notes

1. Manufactured metal bollard hot dip galvanized and painted with either Sika 'Inertol' Icossit Highbuild 5530 paint (or a 2 pack acid etch primer undercoat and gloss). Set bollard into concrete foundations in accordance with manufacturer's instructions.

2. 4 × weathered timber bollard planed finish CCA treated to BS 4072 Part 1 with a dark stained finish 'Tanalized' or similar and approved.

3. Adjacent hard surface.
4. Foundation of C:20:P mix concrete. Final dimensions in accordance with manufacturer's instructions.

General notes
- All dimensions in millimetres
- Do not scale from this drawing

traffic and prevent vehicles using pedestrian areas for access or parking. They can be of timber or metal and are usually 100–200 mm square set into a concrete foundation, with 600–800 mm above ground and 500–700 mm below ground (see Figure 3.15).

Rural areas have a higher proportion of timber fencing and bollards than urban areas, where metal is more usual. Brickwork can be appropriate to both urban

FIGURE 3.15   METAL AND TIMBER BOLLARDS

and rural contexts according to the style of brickwork, which can be rustic or neat accordingly. Metalwork and timberwork can be used above brickwork, although this type of detail is most usually for urban or suburban usage.

Vehicular circulation

Vehicular circulation is an important and extensive subject and highway civil engineers are employed in county and borough councils throughout Britain to advise and make policies on highway design. It is possible therefore to obtain information regarding the design of entrances, crossings, drives, roads and pavements, footpaths and car parks by contacting local highway departments, who after all have to approve such design work before it can be implemented.

The Radburn system

Until the 1970s, the general principles of highway design were those established in Radburn, New Jersey and designed by Stein and Wright in 1927. These principles essentially centred on the separation of pedestrian and vehicular traffic; the vehicular traffic usually taking priority over the pedestrian. The Radburn plan proposed wide roads with as few bends as possible and good sight lines. Pavements would preferably be separated from the road by a grass verge and, in any event, be clearly defined and protected by a high kerb (see Figure 3.16).

FIGURE 3.16 RADBURN SYSTEM OF PEDESTRIAN AND VEHICULAR CIRCULATION

The kerb separation between vehicular and pedestrian zones has become so familiar that it is subliminally understood by most people that the road is the sole domain of the car and the pavement the sole domain of the pedestrian. At first sight the logic of this separation cannot be faulted. It is easy to teach road safety to children by pointing to the kerb, being such a clear division between safe and hazardous areas. However, a new approach was put forward in the 1970s, when it was argued by some that the Radburn approach encouraged fast-moving traffic, even in residential zones where children were often playing in the road. It was argued that some children will always play in residential roads and that in such areas the pedestrians should perhaps take priority over vehicles.

The combined system

For residential areas, especially on roads that are used principally for access, a system was devised during the early 1970s with a combined surface for both pedestrians and vehicles, but which appears to give priority to the pedestrian (see Figure 3.17). By careful control of curves and sight lines and with use of small unit paving materials, cars were successfully made to feel uneasy and speeds could be substantially reduced without external controls.

FIGURE 3.17 THE COMBINED SYSTEM OF PEDESTRIAN AND VEHICULAR CIRCULATION

Such measures have now collectively become accepted practice in highway design for residential access-only roads. Visually, the use of just one material in such areas creates a greater sense of unity. There is also a greater simplicity too, which is often more attractive. The single material for the entire traffic area (pedestrian and vehicular traffic) creates a courtyard appearance, which has a less municipal and impersonal appearance than kerbs, tarmac roads and slab pavements. This new sensitive approach is certainly more appropriate for residential areas.

Contemporary practice

Most housing developers follow the modern conventions of highway design in the layout of their developments, set out in the guidelines prepared by county highway authorities. The spine roads (sometimes called collector roads) and feeder roads are very much based on the Radburn system, with road widths typically from 6 m to 5 m wide for the feeders. However, the culs-de-sac and drives are reduced to 4.5 m and 4 m wide and there are no roadside pavements here as they are designed to be 'combined surfaces'. This combined surface usually starts with a rumble strip or speed hump, which is followed by the change in surfacing – concrete blocks, paving bricks or tarmac of a contrasting colour to the usual blacktop of the spine or feeder roads. The drives themselves are invariably of unit type paving surface.

Sight lines and junctions

Rigid principles are laid down to control the design of new road junctions in order to ensure that the new development and road layout is safe. The main criterion is the preservation of sight lines so that drivers can see other vehicles coming well in advance. The most common sight line criterion on most new estate road junctions is a clear view of some 50 m from a point 2.5 m behind the white give-way lines. The developer (and the landscape designer) must not put any object, be it house, fence, hedge, tree or shrub, greater than 600 mm tall within this zone of vision (see Figure 3.18). Some authorities do not allow trees or shrubs within this zone at all. However, recent concessions from some highway authorities, after much pressure from district council planning departments, have allowed the use of ground cover shrubs within sight lines, so long as their maximum height will not exceed 600 mm throughout their life, the authority taking the view that such planting is likely to be unpruned. A few authorities allow clear-stemmed trees within vision splays, taking a more

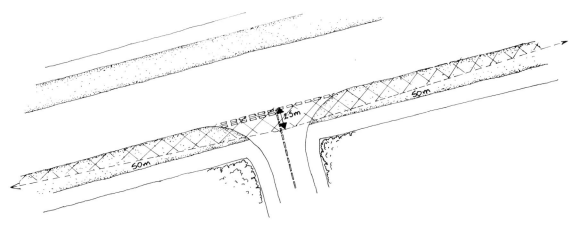

FIGURE 3.18 SIGHT LINES FROM A ROAD JUNCTION. The hatched area is called a 'vision splay'.

 AREA TO BE KEPT FREE OF VISUAL OBSTRUCTIONS

enlightened view that such stems (without branches or leaves) will not interrupt the view.

Pedestrian crossings

The most important factor in siting a pedestrian crossing is to first determine exactly where people desire to cross. The purpose of a crossing is to ensure the safe passage of people across the road. This will not be achieved if the crossing is put in the wrong place and does not get used. Pedestrians will instead cross the road elsewhere, usually the shortest distance between their original position and their destination.

If the point where people most desire to cross is not safe, then the crossing must be as close to the desired spot as possible. Reinforcement measures will be needed, such as guard railings, to enforce the safe use of the crossing. These railings (which must conform to the relevant [British] standards) will stop most people crossing where it is not safe.

The dimensions of a crossing will depend upon whether it is a pelican crossing (controlled by traffic lights) or a zebra crossing (complete with Belisha beacons). These figures can be obtained from the highways authority.

Turning heads

At the end of all 'No Through Roads' the design guides dictate that you should have a suitable turning head to ensure that vehicles, including large service vehicles such as refuse lorries, fire engines, delivery and removals lorries and so on, can turn around easily.

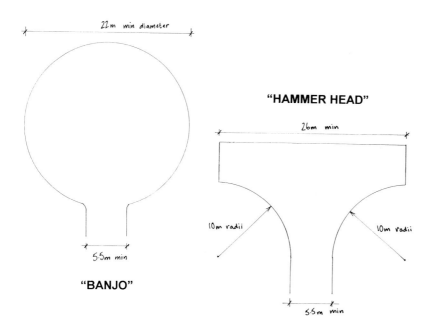

FIGURE 3.19 ROAD TURNING HEADS

There are many forms of turning head (see Figure 3.19). The 'frying pan' or 'banjo' style is a full circle, sufficient to encompass the turning circle of service vehicles (radius minimum = 8.5 m for a refuse vehicle). This is wasteful of space and sometimes harshly creates a sea of hard surfacing. The 'double frying pan' comprises two smaller circles fused together like a dividing cell. This allows smaller vehicles to turn full circle, while larger ones can back into the next circle in order to make a three-point turn.

The 'hammer head' is exactly that, with right-hand and left-hand forks which allow vehicles to reverse into the opposite fork before driving away forwards. This form is functional but dull in appearance and some cars park in the ends, rendering them useless. There is a curved version of the hammer head, giving a more 'Y' shaped appearance, and another 'L' shaped version. While these variants can usefully fit a range of shapes, they encounter the same problem.

Traffic calming

Traffic calming is a comparatively new expression for the appliance of some of the principles of the combined vehicular system to faster roads within residential areas. In places where pedestrians, particularly children, are frequently vulnerable to fast-moving traffic, for example near to a school, the use of pinch points, lane closures, rumble strips and speed ramps and humps have a dramatic effect on reducing traffic speed. Such measures may be irritating for the motorist and sometimes very uncomfortable for the bus passenger and there have been reports that they are causing damage to ambulances and their contents.

One side effect of traffic calming measures is that traffic can migrate elsewhere, solving the immediate problem but sometimes making someone else's much worse. A further problem raised by residents is that the slower traffic increases the amount of car exhaust, while impatient motorists 'rev' their engines at chicanes. Generally, however, these schemes have proved to be very effective for their targeted purposes, particularly in high priority areas such as close to schools.

Car parking

Car parking is an emotive issue. Parked cars are unsightly and it is a nuisance for residents if most of the cars parked in their street are not theirs but instead belong to commuters or tourists. In addition, the large area of paving required for parking has been blamed for causing flooding, because the ground that used to absorb rain-water like a sponge now sheds it into storm drains.

The most space-efficient means of parking many cars is to use the right-angled method, which comprises a row of spaces 2.4–2.5 m wide and 4.8–5.2 m long, with an access and turning lane 6.1 m wide. Such use of space looks dull and is best broken up by an occasional bay for planting. These bays will need protection from the cars with a high kerb, trip rail, bollards or a combination of all three.

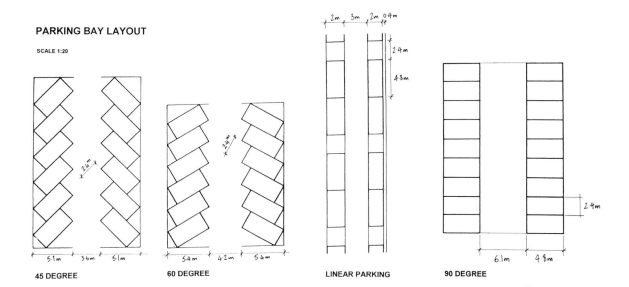

FIGURE 3.20 PARKING BAY LAYOUTS

Alternatively, angled parking can be used (see Figure 3.20), with the spaces set at 30/60 degrees. The access lane can be reduced to 4.2 m for 60 degree parking (and 3.6 m for 45 degree parking) but the method is applicable only to a one-way system and the fact that both an entrance and exit is required usually means that it is a little less space efficient in small areas. Having said that, the pattern is more comfortable to use than the right-angled layout.

Where space does not permit either of these methods, along an existing road or street, a linear parking system is required. The formal layout for this scheme is just a more elaborate method of typical unrestrained street parking but the formal road markings provide for more comfortable exits from and entrances to the spaces. This is achieved by marking out the actual parking bays in pairs, interspersed with two gaps 2.4 m long for access and exit.

Pedestrian circulation

Paths and desire lines

In most situations, people will try to get from one place to another using the shortest route. If you were to plot two places on a map, 'A' (your house) and 'B' (the pub), then the line of desire between these two points would be a straight line (at least for the journey there) (see Figure 3.21). In reality the route may not be straight (as a crow would fly) and all sorts of terrestrial objects may obstruct this desire line, such as houses, trees, fences and so on. The problem for the designer is in accurately assessing these desire lines to accommodate them as far as possible, so saving the users' time, easing their frustration at detours and generally preventing corner-cutting and the problems this causes.

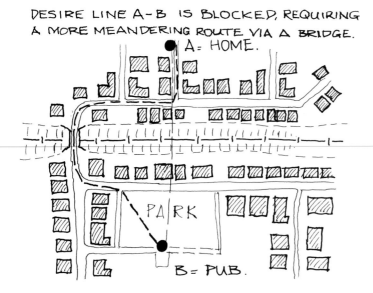

DESIRE LINE A-B IS BLOCKED, REQUIRING A MORE MEANDERING ROUTE VIA A BRIDGE.
A = HOME.
PARK
B = PUB.

Figure 3.21 DESIRE LINES. A desire line is a foreseeable route that people will want to make in a new landscape – an existing path would be termed a route. It is always wise to cater for desire lines as it is hard to block them successfully.

Unfortunately, paths often take a more indirect route than the desire line would suggest, sometimes being laid in straight lines with crude right-angled corners where a change in direction is required. The commonly witnessed result of such routes will be continuous corner-cutting, so that any grass is quickly worn into compacted mud, often filled with rain-water that can no longer drain away. Planting beds will be trampled (even if prickly specimens are used) and fences can also be damaged by corner-cutting.

It may be that a curve would have been the shortest possible way of negotiating an obstacle and would have avoided the problems altogether. However, in many cases even curved paths will cause straight-line short cuts to be made, with the attendant damage to grass and shrubs. This does not mean that paths cannot be curved at all, nor does it mean that all paths must force brutal 'Roman-road'-type straight lines through the site. For a path to succeed in aesthetic and functional terms due consideration and design time are required.

The critical curve

People will follow a curved path so long as the curves are not so severe that they encourage them to leave the path and cut across the unmade ground in between. The maximum extent to which a path can be curved without corner-cutting occurring is called the 'critical curve'. It is important to understand that in perspective a curve will appear far more exaggerated than may be suggested by the drawn plan. That is to say, the effect of foreshortening emphasizes the curved appearance of a wavy path from the perspective (or eye) of the users. It is easy therefore to fall into the trap of making the curve too weak and insipid. Bold curves will always look more attractive than weak, ill-defined ones. To get this right demands a balancing of all these influences to produce the desired 'critical curve' (see Figure 3.22).

Articulation

Articulation is another way of breaking up straight lines. This is the term given to the design of staggered or castellated edged paths, most often seen with stone or concrete slab paths (see Figure 2.13, p. 27). By ensuring that each succeeding row of slabs is out of alignment with the previous ones, an interesting irregular edge can be achieved. This can be manipulated to navigate corners or meanders, or simply to liven up a straight path.

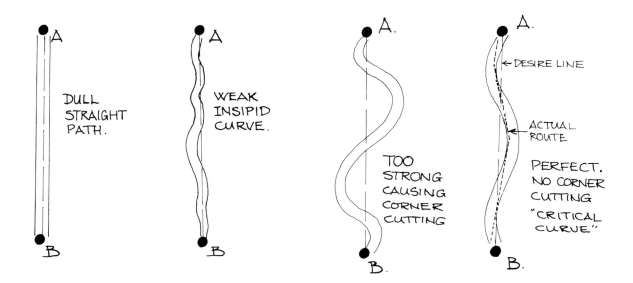

Blocking desire lines

Desire lines may occasionally require blocking. This may be for strategic reasons, to ensure that access is made from one direction only, for example to regulate traffic flow. Sometimes desire lines might be blocked simply to make the best use of the available space, to create mystery by concealing one space from the preceding one or to ensure that a series of spaces (or features) are viewed in a particular sequence and so on. Deliberate blocking of desire lines is, however, a dangerous game. The block will have to be insurmountable if in the public domain, to prevent its failure and the damage that often results. Planting alone is never enough, not even with the use of 'deterrent' plants. The desire line will be used despite the plants and a worn route will become established before the plants ever have time to grow. Where blockage of a desire line is deemed necessary, fencing should be used to reinforce planting. In the controlled environment of the private garden, however, such reinforcement measures are not so necessary and a more relaxed approach can be taken.

Planting should not rely on a single line of shrubs, however dense and tall. The psychological effect of barrier planting is much stronger when there are successive layers of plants, from ground cover plants at the front to tall structural shrubs at the back of the bed (see Figure 3.23). This planting demands sufficient room (minimum of 2 m) to make it effective. In public spaces planting can be perceived to have a menacing connotation, raising fears of concealed muggers, rapists and so on. For this reason it is essential to preserve sight lines by ensuring where possible that planting is first set back from the path and certainly that planting is layered with low and medium height planting nearer to the path to allow for a maximum angle of view.

This important consideration is once again not so necessary within a private garden, where access is controlled and the concealing value of planting

FIGURE 3.22 CRITICAL CURVES. The four diagrams illustrate how a straight line between two points can be improved by a shallow but strong curved line which will direct people to a certain extent along a more meandering route. Beyond a critical point, however, people will leave the path and cut the corner.

FIGURE 3.23 WIDE AND NARROW PLANTING BEDS. Whilst wide planting beds may present a slightly greater physical barrier, the psychological effect of a succession of plant layers is more important in blocking desire lines or controlling access points. The difficulty of using planting for this purpose is with establishment while the plants are small; temporary fencing such as chestnut paling will be required to prevent trampling.

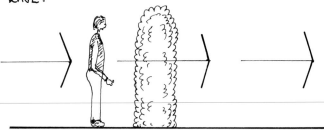

FIGURE 3.24 DESIRE LINES – TREBEY ESTATE. In Trebey Estate, Bow, London, residents complained bitterly about teenagers en route to and from school peering through ground floor kitchen windows, frequent break-ins and general aggravation and noise. To prevent these real problems several metres of paving in front of the houses were given to the ground floor residents and defined by walls and fences. The desire line was blocked by three fences and two blocks of planting yet still a few children persisted in using this route to and from school. Eventually the planting grew sufficiently finally to block the desire line, much to the relief of the residents.

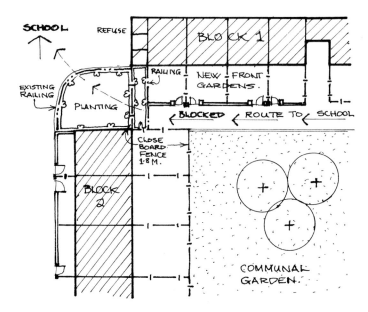

can be employed to its full potential to create both mystery and surprise. When such gardens are known to be safe, the menace of gloomy, shaded spaces and corridors with restricted sight lines can be employed to deliberate effect, to contrast with bright, colourful areas.

In some public spaces desire lines can be very hard to block successfully, even where a total visual screen is employed (such as a 2 m high close board timber fence with planting both in front and behind). On the Trebey Estate, in Bow, London, a desire line ran through the estate (between home and the adjacent school) creating a 'rat-run' for school children past the ground floor windows of flats (see Figure 3.24). The noise, nuisance and some petty crime caused by the 'rat-run' meant that blocking this desire line was a high priority for the Tenants Association.

This apparently straightforward estate improvement scheme proved to be a startlingly difficult job. Eventually a 3 m high fence of unclimbable weld mesh was required. Planting and low fences were provided on each side. To add to the problems, the fence, being over 2 m in height, necessitated a lengthy planning application process, while the route was increasingly being used and becoming more and more vandalized. In the end the scheme did work, much to the residents' relief.

Cross paths

The prevention of corner-cutting is nowhere more frequently found than at cross paths. Here you meet the desire lines between four points 'A', 'B', 'C' and 'D' (see Figure 3.25). It may be that most traffic will run from 'A' to 'B' and from 'C' to 'D'. This would mean that the corner-cutting risk was low. However, there is always likely to be some traffic between 'A' and 'C' and between 'B' and 'D'. The greater this traffic is assessed as being, the greater is the risk of corner-cutting.

If two straight paths cross at 90 degrees, then the corners are likely to become mud baths, unless they are protected. Planting will certainly help to reinforce the junction, but may only be effective in preventing corner-cutting several metres away from the actual junction and will not be enough to prevent cutting of the corners very near to the junction. These corners will require further reinforcement using, at the very least, a knee rail or (preferably) a railing of 900 mm plus, with the planting more densely spaced behind it (see Figure 3.26). The knee rail might be more suited to lower use rural sites, while the railings would be

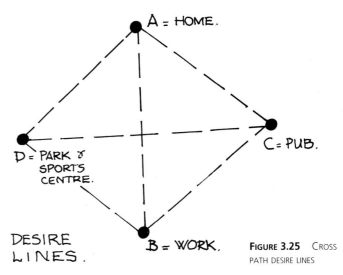

FIGURE 3.25 CROSS PATH DESIRE LINES

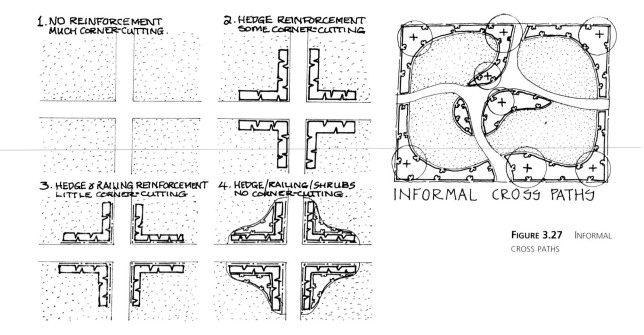

1. NO REINFORCEMENT MUCH CORNER-CUTTING.

2. HEDGE REINFORCEMENT SOME CORNER-CUTTING

3. HEDGE & RAILING REINFORCEMENT LITTLE CORNER-CUTTING.

4. HEDGE/RAILING/SHRUBS NO CORNER-CUTTING.

INFORMAL CROSS PATHS

FIGURE 3.27 INFORMAL CROSS PATHS

FIGURE 3.26 FORMAL CROSS PATHS

appropriate to high use urban sites. In some informal or rural contexts this heavy-handed reinforcement detailing can be avoided by forming the junction at an angle of 60 degrees and slightly staggering it. The approaching paths will need to be carefully meandered to ensure that they are in the right position for this solution to work and plenty of room is required (see Figure 3.27). The method is ideal for an informal site such as a country park, whereas formal and urban areas may well demand the expensive, heavy reinforcement approach required at 90 degree junctions.

Pedestrian crossings

Pedestrian crossings have been covered in the vehicular circulation section but clearly it is essential that a pedestrian crossing is located as close as is practicable to the desire line, otherwise the crossing will not be used.

Path width

It is important when designing paths to work out diagrammatically where the desire lines are and then to grade them according to priority, estimating their likely usage, in order to help decide the material of construction and the width required for each one. It is an obvious point that the more a route is likely to be used, the wider the paths will need to be. This is to ensure that people can proceed without the need to leave the path every time they pass another person or group. Occasional use may require a path to be no more than a metre wide.

Steady, continuous usage would require a 2 m wide path. Heavy use might demand a 3 m wide path.

The main problem in determining the required width of paths is that usage often fluctuates markedly from time to time. The fickle British climate means that it is probably on only half a dozen or so days a year that sunny weather coincides with the holiday season. For these few days there will be heavy usage of a country park but these few days might not justify the cost of installing a 3 m wide path. However, if this width of path is not provided, the countryside that has attracted the visitors will become steadily eroded and damaged during these peak periods.

One good solution (not often seen) to this fluctuating usage is to lay a path of a solid, durable and relatively expensive surfacing of, say, 1 m wide (such as bitumen macadam) for the normal low level usage, with two further lanes on either side of this path of a cheap, less durable surfacing such as crushed stone, hoggin, Breedon gravel or similar, for times of peak usage. Soil erosion at the edges of the main path is prevented by the provision of a path of sufficient width for peak times, but this path will have cost approximately half that of providing a single surfacing of tarmac for the entire width (and will look better too).

Boundaries

Paths can be bounded by many elements – grass, shrubs, hedges, walls, fences, roads or water. The comfortable width of the path and its attractiveness are determined by these elements on the boundary (see Figure 3.28). Enclosing elements, such as shrubs, will make the path appear narrower and this effect is most noticeable with hedges, walls and fences bounding the path. When these elements are on both sides of the path, the effect can be claustrophobic. The sight of a narrow, dark alleyway is rarely appealing and may appear threatening. For this reason the minimum width of such an enclosed path should be 2 m. Paths through low shrubs or grass can be much narrower and still appear com-

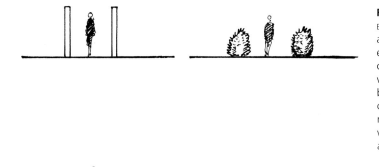

Figure 3.28 PATH ENCLOSURE. The taller and more solid the enclosing boundaries of a footpath, the wider the path should be for it to feel comfortable; that is, a narrow enclosed path will feel threatening and claustrophobic.

fortable. These can be as narrow as 600 mm, or less in a small garden setting. It is important, however, that public open spaces, which are designed for adoption by the local authority, conform to the authority's width and surfacing criteria. Councils may have a minimum width criterion of, say, 1.8 m or 2 m, to ensure the safe passage of wheelchairs and double pushchairs.

Scale

The scale of the space will have a direct effect on the appearance of a path. Large-scale open spaces will require much wider paths so that they appear 'in scale' with the surroundings. Such large spaces have a dwarfing effect and a 1 m wide path in an open heath will seem very narrow to the user, while the same path in a small garden would appear like a motorway.

Changes of level

Changes in level can be achieved by retaining walls, steps, ramps, step ramps, kerbs, terraces and rockeries. These landscape elements are explained in detail below.

Differences in level on a site can be accommodated by sloping the ground. The slope must, however, be shallower than the angle of repose, which is the angle (from horizontal) or slope at which the soil remains stable. This angle is different for every soil type. Clay-gate, a peculiar sandy-clay found near Brentwood in Essex, has a very low angle of repose because it has no structural strength. Most rock, of course, has a very steep angle of repose.

Retaining walls

Changes of level can present problems because sloping ground is not always appropriate to the proposed land use, especially where space is limited, often demanding the maximum possible area of flat ground. In such circumstances a retaining wall will be required (see Figure 3.29). Any retaining wall (even those only 300 mm high) needs to be strong enough to hold back the weight of soil and also water pressure. Water pressure can be minimized by the inclusion of weep holes, which should be of a suitable diameter to avoid clogging. Geotextile should be used to protect the back of the wall and the wall itself should be waterproofed by a bitumen compound like 'Synthapruf' using two to three coats. Sometimes a filter drain can be used such as ICI 'Filtram' to take water to the weep hole levels or to a horizontal land drain at the base of the wall. Behind this membrane an area of granular fill material such as gravel, hardcore or crushed stone can be used to fill the gap between the soil and the wall, which also allows the water to drain to the weep holes.

Battered walls are stronger than vertical ones because some of the weight of soil is removed and the wall sits back against the load. The concrete foundations of retaining walls are an important element because the toe of the footing needs to extend out from the wall much further than for a free-standing wall.

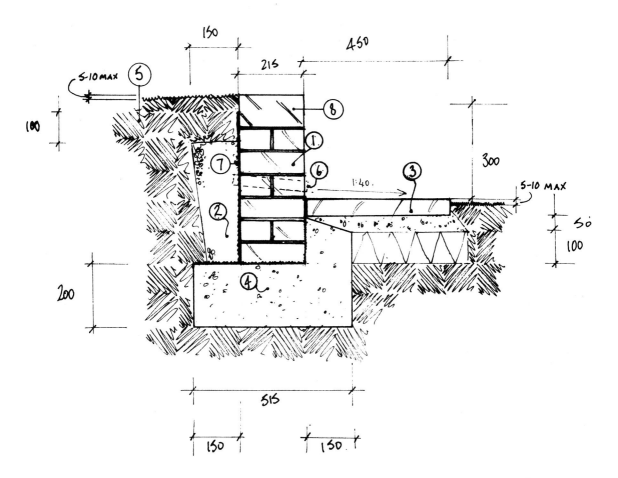

Notes
1. Stock bricks bedded, jointed and bucket-handle pointed in 1:3 mix cement mortar.
2. Backfill of MOT Type 1 granular fill or similar and suitable hardcore placed ensuring no hardcore pieces block or restrict drainage pipe bores.
3. Adjacent surface of 450 mm × 450 mm × 50 mm precast concrete slabs.
4. Concrete strip foundation of C:20:P mix.
5. Adjacent turf grass surface to be maximum of 10 mm below adjacent surface to receive surface water.

6. PVC weep pipe (No. 37) @ 3000 mm centres 75 mm diameter.
7. 2 No. coats of Synthapruf bituminous paint.
8. Brick on edge coping.

General notes
- All dimensions in millimetres
- Do not scale from this drawing

Brick retaining walls require a very thick base. This can narrow towards the top of the wall where the loading is less severe. Common bricks can be used below ground or wherever they cannot be seen to minimize the cost but nevertheless the base may need to be over a metre thick.

A less expensive alternative is to build a retaining wall of dense masonry concrete blocks (breeze blocks) with steel reinforcement bars through the

FIGURE 3.29 BRICK RETAINING WALL

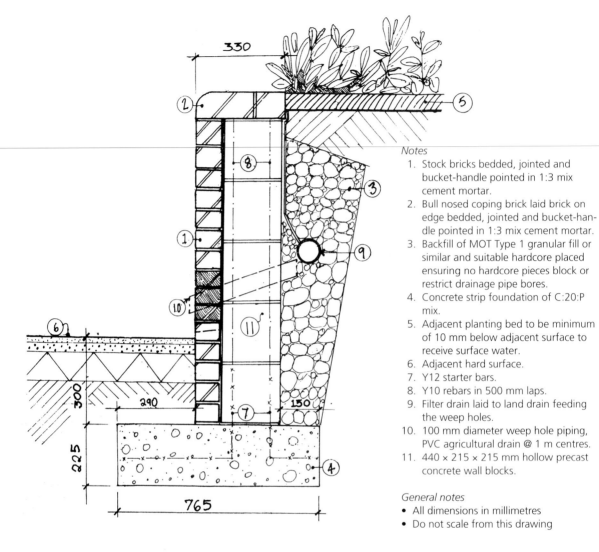

Notes
1. Stock bricks bedded, jointed and bucket-handle pointed in 1:3 mix cement mortar.
2. Bull nosed coping brick laid brick on edge bedded, jointed and bucket-handle pointed in 1:3 mix cement mortar.
3. Backfill of MOT Type 1 granular fill or similar and suitable hardcore placed ensuring no hardcore pieces block or restrict drainage pipe bores.
4. Concrete strip foundation of C:20:P mix.
5. Adjacent planting bed to be minimum of 10 mm below adjacent surface to receive surface water.
6. Adjacent hard surface.
7. Y12 starter bars.
8. Y10 rebars in 500 mm laps.
9. Filter drain laid to land drain feeding the weep holes.
10. 100 mm diameter weep hole piping, PVC agricultural drain @ 1 m centres.
11. 440 × 215 × 215 mm hollow precast concrete wall blocks.

General notes
- All dimensions in millimetres
- Do not scale from this drawing

FIGURE 3.30 BRICK-FACED RETAINING WALL

central holes, which are then filled with lean mix concrete. The face of the wall can then be disguised by a brick skin of facing bricks (see Figure 3.30).

Precise calculations, obtained from a qualified structural engineer, are recommended for retaining walls much above 2 m in height. For walls below 2 m, 'off-the-peg' construction details can be used successfully. Foundations may need to be much deeper than shown to ensure that they are laid on solid ground. This can only be established on site after trial pits have determined the nature of the soil. Construction details should indicate that the contractor should allow in its price for any additional labour and materials, overheads and profit in increasing foundation depths in soft ground until a firm base for these foundations is achieved. This stipulation also applies to free-standing walls, of course.

Geotextiles

The retaining of sloping land can be achieved in many ways. It is important to know the angle of repose of the soil material when any soil embankment is encountered or proposed. It is possible to stabilize sloping ground so that steeper gradients can be achieved than the angle of repose, using a variety of methods and materials. These methods include the use of horizontal geotextile strips in sandwich layers or indeed laid over the face of the slope. Proprietary soil-filled mesh material, such as those produced by Netlon and others, complete with integral grass seed, can be laid over the face of sloping ground. The grass grows and the roots bind the material to the slope, forming a dense, stable mat. There are also a variety of fibrous matting materials available, some ready impregnated with seed and fertilizer, which are laid over the surface and do the same job.

The use of geotextiles is particularly appropriate in more rural contexts where there is often sufficient space to allow the ground to slope to some degree. This may even be to the natural angle of repose for the particular soil conditions and then grass or other plants may be employed to minimize soil erosion.

Timber and concrete structures

For more vertical retention, timber can be used, comprising 50 mm diameter

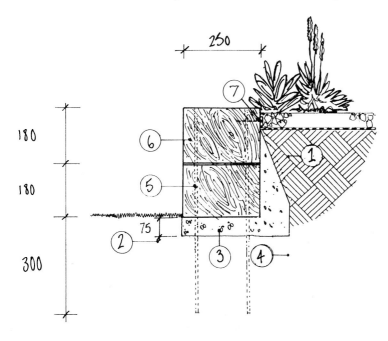

FIGURE 3.31 SECTION DETAIL OF SLEEPER RETAINING WALL

Notes
1. Prepared topsoil, including treatment of sheet mulch and bark chip. Final levels a maximum of 10 mm below upper surface of sleeper retaining wall.
2. Adjacent turf grass surface.
3. Infill of well-consolidated washed pea shingle.
4. Well-consolidated subgrade.
5. Steel reinforcing rods cut to 660 mm in length and driven through holes to 150 mm below ground level ensuring tops of rods are sunk below surface of railway sleepers.
6. Reclaimed railway sleepers laid to form soil retaining wall to give a flat neat face along the entire length facing outwards.
7. Galvanized flat head nails of suitable length driven through sheet mulch into railway sleeper as detailed at 1 m centres.

General notes
- All dimensions in millimetres
- Do not scale from this drawing

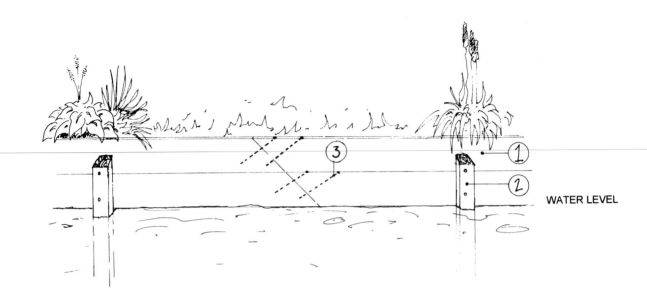

WATER LEVEL

Notes

1. Sawn softwood boarding 38 × 200 mm as retaining board angle jointed with 2 No. 150 mm galvanized nail fixings per joint per board.
2. 50 × 50 × 600 mm stake with once weathered top driven into ground at centre of gravel board and fixed to gravel board with 50 mm galvanized nail fixing.
3. Galvanized nails 150 mm length.

General notes

- All dimensions in millimetres
- Do not scale from this drawing
- All timber to be CCA treated to BS 4072 Part 1

FIGURE 3.32 TIMBER RETAINING STRUCTURE

pressure-treated stakes with boards set behind them to hold back the soil, which again can be set at an angle. Plants can then be grown into the crack and joints. A more sophisticated form of this principle are proprietary makes of interlocking timber or concrete sections, usually battered, which leave slots of exposed soil for planting trailing shrubs. These are effective but are usually expensive and poor to look at until the plants establish, which they often fail to do – especially on south facing slopes which are prone to extremes of temperature and drying out. Nevertheless, the system can be used effectively to stabilize very considerable changes in level (tens of metres) (see Figures 3.31 and 3.32).

Gabbions and piling

Gabbions (metal cages) filled with stones are a common method of stabilizing riverbanks, where plenty of stone is available. Where it is not, river authority engineers use steel sheet piling, particularly where wash from boats is likely to affect the banks. This system has caused some problems – farm, domestic and wild animals that fall into the rivers and canals are unable to climb out, and then drown. This has prompted wildlife activists to request ramps at intervals to allow animals to swim to shore, but the expense has so far proved prohibitive.

Sandbags

Sandbags, especially when filled with dry lean mix, can be used to create battered retaining walls (see Figure 3.33). These bags solidify into concrete and soon green over with moss, especially if north facing or near water.

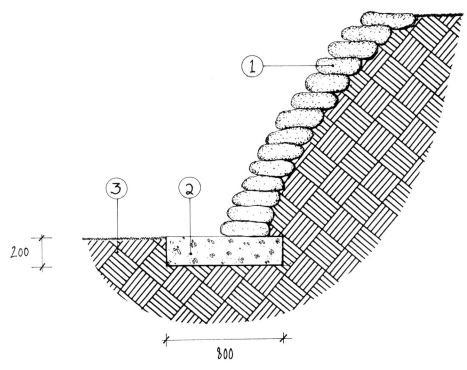

200

800

Notes
1. Stacked sandbags filled with sand-cement lean mix.
2. C:20:P mix concrete strip foundation 200 mm depth.
3. Adjacent topsoil.

General notes
- All dimensions in millimetres
- Do not scale from this drawing

FIGURE 3.33 BATTERED BAG RETAINING WALL

Steps

Where access is required between two different levels a ramp, stepped ramp or, particularly where space is limited, steps can be used. Steps are the traditional approach to accessing different levels, but increasing awareness of people with disabilities has made designers use ramps whenever possible or preferably combine the two. The main incentive for using steps in these more enlightened times is aesthetic, for example to create a grand statement.

Steps can be used to express great strength and authority with their smooth architectural lines where such steps are given a grand setting and sufficient scale. Steps are often used in association with civic buildings and those of financial and government institutions. In contrast, narrow, irregular steps can

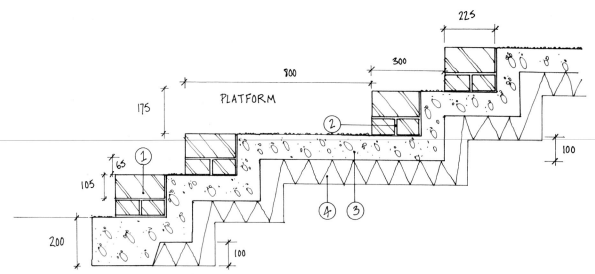

Notes

1. Ockley ATR selected second stock brick as step laid brick on edge over stretcher course.
2. Bedding, flush pointing and jointing in 1:3 mix cement mortar.
3. Step foundation of C:20:P mix concrete exposed aggregate on platforms by allowing concrete to set for 15 minutes after laying and brushing with a stiff broom to expose aggregate.

4. MOT Type 1 granular fill or similar and approved clean hardcore well consolidated to 100 mm thickness over well-consolidated subgrade.

General notes
- All dimensions in millimetres
- Do not scale from this drawing

FIGURE 3.34 BRICK AND EXPOSED AGGREGATE STEPS WITH PLATFORM

be very intriguing in a garden setting, especially when the destination is concealed.

In a rural context, steps can be made from crude lumps of stone or log risers and treads made from crushed stone on a well-rammed and firm base. This simple detail can soon become dangerous to use unless the treads are well consolidated, as they otherwise soon become dished and uneven. The logs are held in place by 50 mm timber stakes at the sides of the path. These must be placed far enough to the sides of the path to ensure that the users' feet do not land on them and throw them forward if walking down the steps. The logs will eventually rot, though pressure-treated timber or railway sleepers will endure for far longer than ordinary logs.

Risers should be within the range 125–200 mm for comfort and treads 200–350 mm. Outside these dimensions, steps can be very uncomfortable to walk up, particularly with a low riser and a very wide tread that is just longer than a normal pace. When the tread is widened to a platform, then it is comfortable again and such platforms should be 800–1200 mm (see Figure 3.34).

Generally, steps will be constructed on a base of concrete, the profile of the steps created with timber form work (see Figure 3.35). On this concrete base many different materials can be used for the risers and treads. The more usual possibilities are listed on page 78.

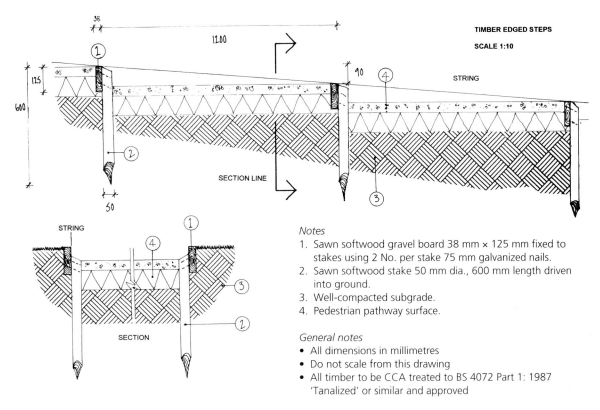

Notes
1. Sawn softwood gravel board 38 mm × 125 mm fixed to stakes using 2 No. per stake 75 mm galvanized nails.
2. Sawn softwood stake 50 mm dia., 600 mm length driven into ground.
3. Well-compacted subgrade.
4. Pedestrian pathway surface.

General notes
• All dimensions in millimetres
• Do not scale from this drawing
• All timber to be CCA treated to BS 4072 Part 1: 1987 'Tanalized' or similar and approved

FIGURE 3.35   TIMBER-EDGED STEPS

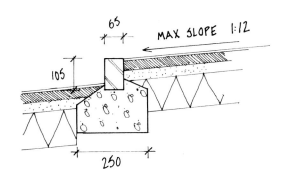

BRICK ON EDGE AS RISER

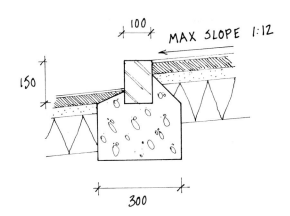

GRANITE / CONCRETE KERB AS RISER.

FIGURE 3.36   STEP RAMP: RISER DETAILS

1. Brick on edge risers on one stretcher bond course of brick for the risers with a concrete tread, possibly exposed aggregate finish. The tread will need to be a minimum of 100 mm depth over 100 mm of well-consolidated and blinded hardcore or granular fell (see Figure 3.36).
2. Brick risers with pcc slab or stone treads (see Figure 3.37).
3. Brick treads and risers (see Figure 3.38).
4. In situ concrete risers and base plus precast concrete slab treads (see Figure 3.39).

Long flights of steps (of more than 50 steps) are best interrupted with a platform so that a single flight has no more than 10 steps.

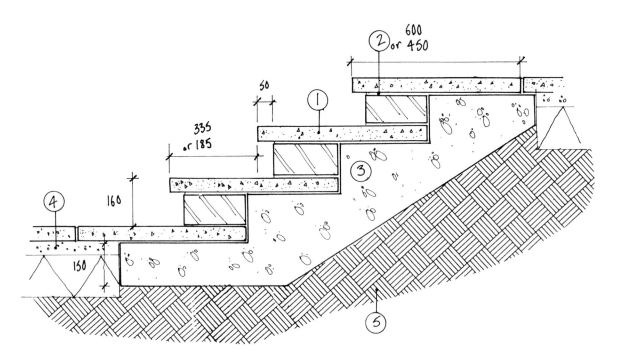

Notes
1. Nominal size precast concrete paving slabs as treads laid over brick on edge soldier course as step.
2. Bedding, flush pointing and jointing of 1:3 mix cement mortar.
3. Step foundations of C:20:P mix concrete.
4. Adjacent paving surface.
5. Well-consolidated subgrade.

General notes
- All dimensions in millimetres
- Do not scale from this drawing

FIGURE 3.37 BRICK AND SLAB STEPS

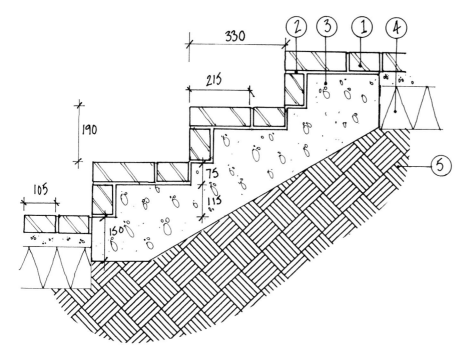

Notes
1. Ockley ATR seconds laid brick on flat with soldier course backed on to stretcher course to form brick treads and steps.
2. Bedding, flush pointing and jointing of 1:3 mix cement mortar.
3. Step foundations of C:20:P mix concrete.
4. Adjacent brick paving surface.
5. Well-consolidated sub-grade.

General notes
- All dimensions in millimetres
- Do not scale from this drawing

FIGURE 3.38 BRICK TREADS AND RISERS

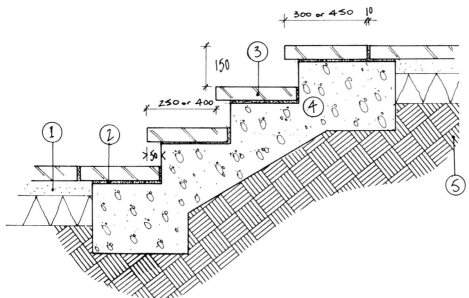

Notes
1. Adjacent paving surface.
2. Bedding, flush pointing and jointing of 1:3 mix cement mortar.
3. Precast concrete paving slabs as treads.
4. Step foundations of C:20:P mix concrete.
5. Well-consolidated sub-grade.

General notes
- All dimensions in millimetres
- Do not scale from this drawing
- All surfaces laid to falls and cross falls as detailed

FIGURE 3.39 IN SITU CONCRETE RISERS AND BASE PLUS PCC SLAB TREADS

Ramps

Ramps must conform to the relevant standards, which dictate that the maximum gradient is 1:12 m. Shallower slopes than this are obviously easier for wheelchair users, although very shallow gradients will require longer to get from A to B and will cost more. The surface of the ramp should be smooth to create less friction and to be more comfortable for wheelchairs. High quality bitumen macadam is the best surface in this respect. Steps and ramps can be positioned together as combined units with a separating retaining wall (see Figure 3.40).

FIGURE 3.40   RAMP

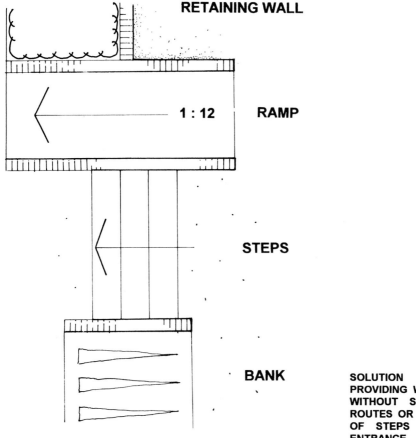

RETAINING WALL

1 : 12   RAMP

STEPS

BANK   SOLUTION TO PROBLEM OF PROVIDING WHEELCHAIR ACCESS WITHOUT SEPARATING ACCESS ROUTES OR LOSING THE EFFECT OF STEPS ANNOUNCING AN ENTRANCE

Step ramps

Very long steep slopes can be accommodated using a step ramp, which at least allows wheelchair access if assisted (see Figure 3.41). A 1:12 m ramped section is interspersed with a low vertical riser, usually of 100 mm maximum. Stone or concrete kerbs can be used for the risers or bricks laid on edge or flat.

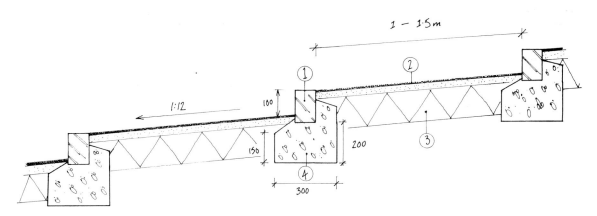

Notes
1. Riser unit.
2. Bitmac surface material.
3. 150 mm depth hardcore.
4. C:20:P mix concrete foundation.

General notes
- All dimensions in millimetres
- Do not scale from this drawing

FIGURE 3.41 STEP RAMP: BITUMEN MACADAM WITH BRICK OR GRANITE KERB RISER

Kerbs and edgings

Kerbs and edgings are used for both pedestrian and vehicular pavements, both rigid and flexible in their construction. Kerbs and edgings are especially important for flexible surfaces as these will suffer the greatest movement with wear (see Figure 3.42).

FIGURE 3.42 BRICK KERB USED AS BARK PIT EDGE FOR PLAY AREA, MORETON HALL ESTATE, BURY ST EDMUNDS

FIGURE 3.43 PRECAST
CONCRETE KERB DETAILS

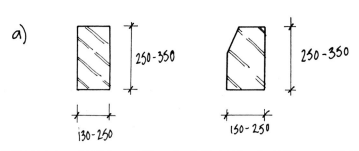

a)

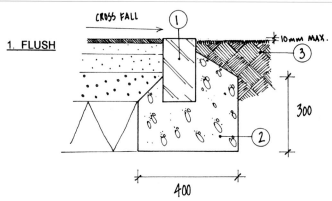

1. FLUSH

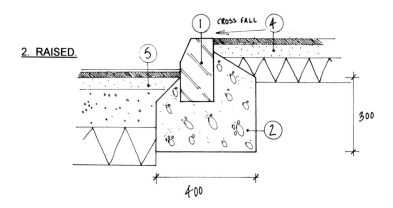

2. RAISED.

Notes
1. PCC kerb unit.
2. C:20:P mix concrete foundation.
3. Grass 5–10 mm max. below paving surface to receive surface water.
4. Pedestrian grade bitmac surfacing.
5. Vehicular grade bitmac surfacing.

General notes
• All dimensions in millimetres
• Do not scale from this drawing

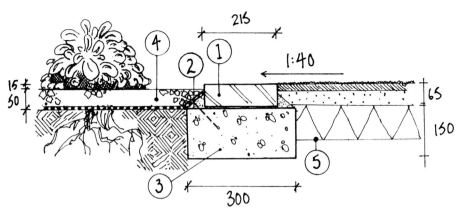

Notes

1. Edging to be 215 mm wide brick on flat using Ockley ATR seconds selected for paving laid brick on flat bedded, flush pointed and 10 mm jointed on 1:3 mix cement mortar.
2. Cement mortar 1:3 mix.
3. Strip foundation 150 mm × 300 mm C:20:P mix concrete.
4. Adjacent planting bed including bark mulch over sheet mulch. Final levels to be 15 mm below surface of adjacent brick edging.
5. Body of adjacent pathway.

General notes

- All dimensions in millimetres
- Do not scale from this drawing
- All surfaces to be laid to falls and cross falls

FIGURE 3.44 BRICK EDGING BETWEEN PATHWAY AND PLANTING BED

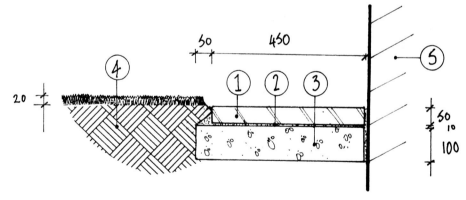

Notes

1. 450 mm × 450 mm × 50 mm precast concrete slabs bedded, jointed, haunched and flush pointed with 1:3 mix cement mortar.
2. 1:3 mix cement mortar.
3. 500 mm × 100 mm strip foundation of C:20:P mix concrete.
4. Adjacent turf grass surface.
5. Adjacent building.

General notes

- All dimensions in millimetres
- Do not scale from this drawing

FIGURE 3.45 MOWING MARGIN NEXT TO BUILDING

Kerbs may be laid flush to road surfaces or raised to afford a definite change of level between pedestrian and vehicular traffic (see Figure 3.43). This provides a 'safe' zone for pedestrians and prevents cars parking on pedestrian pavements. Edgings can also be laid either proud or flush to the paving. Flush edgings allow surface water to drain to grass areas or flower beds adjacent to the paths whilst raised edgings will cause a requirement for surface water drainage (see Figures 3.44 and 3.45). Raised edgings tend to use bull nosed edgings while flush edgings always use the flat-topped stone.

Kerbs may be constructed from brick, stone or their cheaper alternative, precast concrete blocks. Concrete is the most common material for edging and kerbs and is available in precast units of varying sizes and grades, all of which must conform to [British] standard regulated sizes. They are mostly rectangular in shape, with or without a chamfered-edged face, though some of the more expensive units (which mimic natural stone products) may come in a greater variety of shapes and sizes (see Figures 3.46–3.51). Brick or block kerbs are often constructed using soldier courses of 200 × 100 × 65 clay or concrete paviors, or the traditional brick 215 × 105 × 65 sometimes battered at a 30 or 45 degree angle (see Figures 3.52 and 3.53).

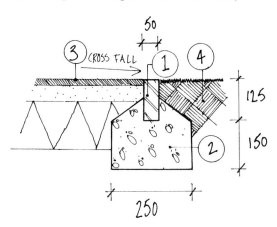

Notes
1. 125 mm × 50 mm precast concrete edging.
2. 150 mm × 250 mm strip foundation C:20:P mix concrete.
3. Adjacent road surface.
4. Adjacent turf grass 5–10 mm below paved surface to receive surface water.

General notes
- All dimensions in millimetres
- Do not scale from this drawing
- All surfaces to be laid to falls and cross falls as detailed

FIGURE 3.46 FLUSH PRECAST CONCRETE EDGING

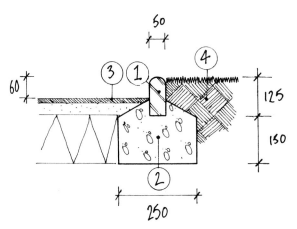

Notes
1. 125 mm × 50 mm precast concrete edging.
2. 150 mm × 250 mm strip foundation C:20:P mix concrete.
3. Adjacent road surface.
4. Adjacent turf grass.

General notes
- All dimensions in millimetres
- Do not scale from this drawing
- All surfaces to be laid to falls and cross falls as detailed

FIGURE 3.47 PRECAST CONCRETE EDGING WITH CHANGE IN LEVELS

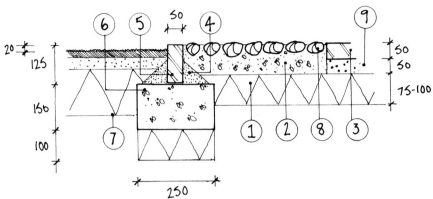

Notes

1. MOT Type 1 granular fill or similar and approved clean hardcore well consolidated to 75–100 mm depth to falls and cross falls as detailed.
2. 100 mm thickness C:20:P mix concrete.
3. 450 mm × 450 mm × 50 mm precast concrete slabs bedded, jointed, haunched and flush pointed with 1:3 mix cement mortar.
4. 1:3 mix cement mortar pointing, bedding and haunching.
5. 125 mm × 50 mm precast concrete edging.

6. 150 mm × 250 mm strip foundation C:20:P mix concrete.
7. Adjacent road surface.
8. Scottish beach cobbles laid with axes parallel and flat side uppermost.
9. 50 mm thickness consolidated sharp sand.

General notes
- All dimensions in millimetres
- Do not scale from this drawing
- All surfaces to be laid to falls and cross falls as detailed

FIGURE 3.48 JUNCTION BETWEEN PRECAST CONCRETE PAVING AND DRIVE WITH COBBLE BORDER

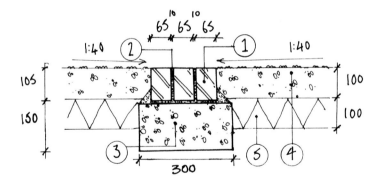

Notes

1. Edging to be 3 No. Ockley ATR seconds selected for paving laid brick on edge bedded, flush pointed and 10 mm jointed on 1:3 mix cement mortar.
2. 1:3 mix cement mortar, pointing, jointing, bedding and haunching.
3. 150 mm × 300 mm strip foundation C:20:P mix concrete.
4. Adjacent paving surface of 100 mm thickness exposed aggregate C:20:P mix concrete laid to falls of 1:40.

5. MOT Type 1 granular fill or similar and approved clean hardcore well consolidated to 75–100 mm depth to falls and cross falls as detailed.

General notes
- All dimensions in millimetres
- Do not scale from this drawing
- All surfaces to be laid to falls and cross falls

FIGURE 3.49 BRICK EDGING BETWEEN PEDESTRIAN GRADE CONCRETE PAVING SURFACES

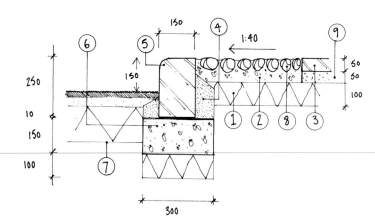

FIGURE 3.50 JUNCTION BETWEEN PRECAST CONCRETE PAVING AND ROAD WITH COBBLE BORDER

Notes

1. MOT Type 1 granular fill or similar and approved clean hardcore well consolidated to 75–100 mm depth to falls and cross falls as detailed.
2. 100 mm thickness C:20:P mix concrete.
3. 450 mm × 450 mm × 50 mm precast concrete slabs bedded, jointed, haunched and flush pointed with 1:3 mix cement mortar.
4. 1:3 mix cement mortar, pointing, jointing, bedding and haunching.
5. 150 mm × 250 mm precast concrete kerb unit.
6. 150 mm × 300 mm strip foundation C:20:P mix concrete.
7. Adjacent road surface.
8. Scottish beach cobbles laid with axes parallel and flat side uppermost.
9. 50 mm thickness consolidated sharp sand.

General notes

- All dimensions in millimetres
- Do not scale from this drawing
- All surfaces to be laid to falls and cross falls as detailed

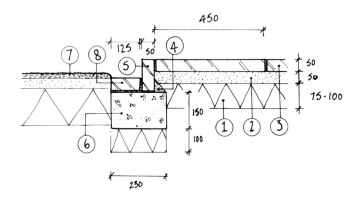

FIGURE 3.51 JUNCTION BETWEEN PRECAST CONCRETE PAVING AND ROAD

Notes

1. MOT Type 1 granular fill or similar and approved clean hardcore, well consolidated to 75–100 mm depth to falls and cross falls as detailed.
2. 50 mm thickness sharp sand.
3. 450 mm × 450 mm × 50 mm precast concrete slabs laid jointed and flush pointed with 1:3 mix cement mortar.
4. 1:3 mix cement mortar as haunch.
5. 125 mm × 50 mm precast concrete edging unit.
6. 150 mm × 230 mm strip foundation of C:20:P mix concrete.
7. Adjacent road surface.
8. 125 mm × 50 mm precast concrete channel unit.

General notes

- All dimensions in millimetres
- Do not scale from this drawing

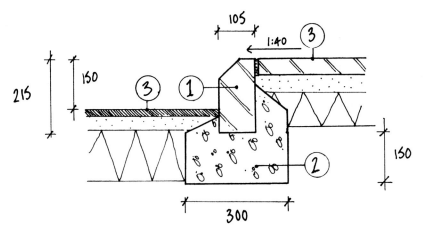

Notes

1. Edging to be Ockley ATR seconds, or single cant stock, laid brick on end, soldier bond bedded, flush pointed and 10 mm jointed on 1:3 mix cement mortar.
2. 150 mm × 300 mm strip foundation C:20:P mix concrete.
3. Adjacent paving surface.

General notes

- All dimensions in millimetres
- Do not scale from this drawing
- All surfaces to be laid to falls and cross falls as detailed

FIGURE 3.52 SOLDIER COURSE BRICK KERB

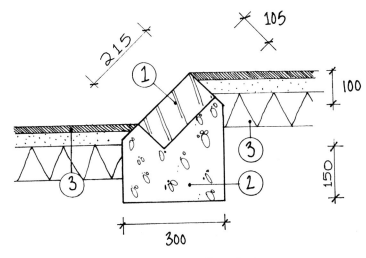

Notes

1. Edging to be stock brick laid at 45 degrees away from road surface bedded, flush pointed and 10 mm jointed on 1:3 mix cement mortar.
2. 150 mm × 300 mm strip foundation C:20:P mix concrete.
3. Adjacent paving surface.

General notes

- All dimensions in millimetres
- Do not scale from this drawing
- All surfaces to be laid to falls and cross falls as detailed

FIGURE 3.53 BATTERED SOLDIER COURSE BRICK KERB

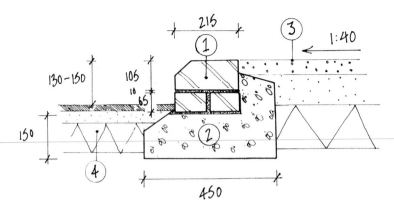

Notes
1. Kerb to be single cant brick laid brick on edge over stretcher course; all bedded, flush pointed and 10 mm jointed on 1:3 mix cement mortar.
2. 150 mm × 450 mm strip foundation C:20:P mix concrete haunched around brickwork.
3. Adjacent pedestrian paving surface.
4. Adjacent vehicular surface.

General notes
- All dimensions in millimetres
- Do not scale from this drawing
- All surfaces to be laid to falls and cross falls as detailed

FIGURE 3.54   BRICK KERB

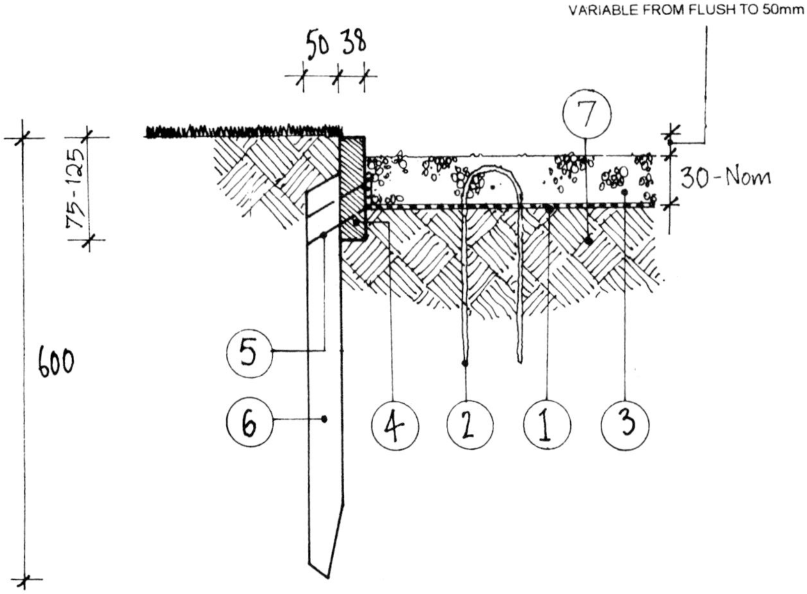

VARIABLE FROM FLUSH TO 50mm

Notes
1. Hy-Tex '18 Ground Cover' – agrotextile sheeting.
2. Wire hoop pegs at 500 mm centres to fasten sheeting.
3. 30 mm – nominal depth of washed pea shingle.
4. Sawn softwood gravel board 38 mm × 75–125 mm. This can be reduced to 33 mm × 75 mm for bending around tighter curves; cut along back with a saw and soak to assist forming a smooth curve.
5. Galvanized nails 70 mm, 2 No. per stake driven through gravel board into stake.
6. Softwood stakes 50 mm dia., 600 mm length at 1.5 m centres; closer if required at junctions or on curves.
7. Well-consolidated subgrade.

General notes
- All dimensions in millimetres
- Do not scale from this drawing
- All timber to be CCA treated to BS 4072 Part 1: 1987 'Tanalized' or similar and approved

FIGURE 3.55   JUNCTION OF GRAVEL PATH AND SOFT WORKS USING TIMBER EDGING

An alternative brick kerb involves the use of a single special cant brick laid on edge over a course of bricks laid flat, which gives a very attractive appearance, although it is not cheap (see Figure 3.54).

The cheapest form of edging is timber board edging, which is mostly specified as sawn softwood, pressure-treated gravel board 25–38 mm wide × 75–125 mm deep flush with surfacing and held in place with 600 mm long pressure-treated 50 mm diameter timber stakes driven into the ground and nailed through into the board using galvanized nails (see Figure 3.55). Corners can be turned at shallow radii (3 m or larger) by bending the boards and cutting the back of the boards a third through and soaking them to ease the bending of the wood, without the cuts being deep enough to cause a fracture. For very tight radii plywood can be used, though this is less durable.

An extension of this board and stake principle can be used to terrace slopes and even to line ditches and small streams. Up to a metre of soil could be retained this way and with modern timber treatment softwood boards can last 10 years or more. Such treatment includes CCA treatment to BS 4072 Part 1 (of which proprietary brand names exist such as 'Tanalized').

Terracing

Terracing is the term used to create a series of flat planes or surfaces with a short drop to each successive level relative to the overall fall or land retained. A series of gardens or spaces for other purposes can be created avoiding the requirement for large and expensive retaining structures. Terracing does, however, take up a lot of room, and necessarily breaks up a large area into smaller spatial units, which may or may not be appropriate to the context or use intended. Terraces employ a series of retaining structures of modest height and these could be constructed from any of the methods described above.

Rockeries

Rockeries can be used to retain soil in the same way as terracing. The difference between properly laid stone and badly laid stone is immense. Most garden rockeries are little better than scattered stone. To work, rockeries must be laid to mimic natural rock outcrops, on a miniature scale, because they are miniatures of real rock outcrops. It is essential to site them where there are no large-scale features, including trees, which can act as a reference point for true scale, as these will destroy the intended illusion.

The rocks must be laid flat with the grain of the rock horizontal (see Figure 3.56). All the rocks laid must be true to the pattern or they will appear unnatural and out of place. Care must be taken to turn and manoeuvre the rocks to ensure the grain and face of the rock is in line with others. The rocks must be laid in bands, the largest at the bottom and the smallest at the top to mimic the rock strata found in natural rocky countryside. The steepest faces of rock should be at the bottom; nearer to the top of the rockery the gradient should be much less and may disappear into a shale scree or gravel top where

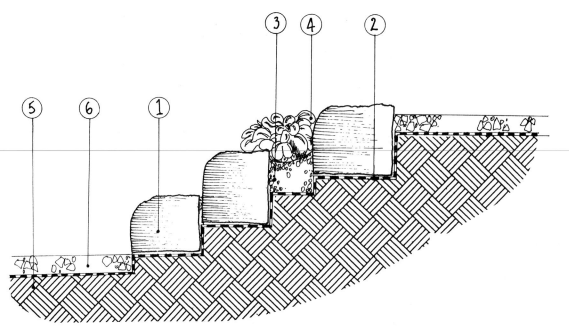

Notes

1. Rockery rocks of nominal size laid with grain of rocks all in same direction to mimic naturally forming strata leaving irregular spaces for specimen rock plants. 200 mm × 200 mm min. suggested type to be weathered, frost resistant sandstone or limestone blocks.
2. Hy-Tex '18' agrotextile sheeting.
3. 50–100 mm depth washed pea shingle laid over sheet mulch surrounding rocks.
4. Rockery plants species planted as detailed. Plant between rocks cutting slits in sheet mulch with scissors. Remove soil in available gaps and backfill with soil/compost mix. Plant container grown stock in gaps firming well in and water in with 10 litres of water per plant. Fold back sheet mulch and replace gravel around base.
5. Graded and profiled topsoil.
6. Adjacent planting bed including bark/sheet mulch treatment..

General note
• Do not scale from this drawing

FIGURE 3.56 SECTION DETAIL ORNAMENTAL ROCKERY

rockery plants can carpet the ground. The stones must overlap the rows in front but not so regularly that they appear in regimented rows like soldiers. The spaces between the rocks are traditionally filled with pea shingle and preferably the whole lot should be laid over a geotextile sheet, such as ICI 'Terram 1000' or Hy-Tex UK's 'Hy-Tex 18', which will prevent the rockery becoming weed-choked.

Paving

Paving design is crucial to the success of the landscape design as it is often one of the most functional elements and as such it must perform well to ensure the success of the landscape design. Good paving design is a synthesis of the optimum combination of the following criteria: context, durability, pattern and cost.

Effective paving design is a product of employing appropriate design principles for the context and then using a good working knowledge of construction technology in the detailing. It is important to understand the character appropriate to the intended usage and then to understand the scale of the site in resolving the paving design, for example the larger the scale the bolder the design should be. From such philosophical beginnings the individual materials can be selected, taking into account aesthetics, cost and durability, and clearly it is sometimes necessary to find a compromise between these three conflicting forces.

Context

Context must be the first consideration in paving design because the budget, pattern and durability criteria alone might well suggest a perfectly good paving that is entirely inappropriate for the setting. For example, a medium cost paving of high durability and simple pattern would suggest pressed precast concrete slabs but these would not be appropriate for a rural or conservation area, where perhaps a bound gravel surface would be more suitable.

Similarly, a busy street will demand a different approach to a quiet parkland setting, although both may be situated in a broadly urban setting. The busy street will be concerned with fast-moving traffic, demanding a simple, bold pattern, highly durable and perhaps with some directional emphasis. In the park a more intricate design is possible, as people will be travelling at a more relaxed speed and will be able to take in and appreciate more complex patterning. Patterns will tend to be static in type rather than directional.

Durability

Choosing a paving that will stand up to the strain of the particular amount of wear and tear at a site is fundamental to the satisfactory functioning of a landscape design. The main differences between hard materials chosen in landscape design occur between pedestrian grade and vehicular grade materials and between natural, waterbound paving products such as Breedon gravel and hoggin, and manmade products such as bricks, blocks, slabs and tarmacadam materials.

Vehicular paving clearly has to be far tougher than pedestrian, to withstand the load of the vehicles. How many times have you seen cracked paving slabs outside shops, where delivery vehicles have mounted the kerb and pedestrian grade slabs have been unable to stand the weight? Standard pedestrian slabs are 50 mm thick, but can be obtained in 70 mm thickness for areas prone to occasional vehicular traffic. Bricks are normally 65 mm thick but can be made 80 mm thick for vehicular areas and edgings of roads. Tarmacadam surfacings and base courses will be made double thickness for vehicular traffic and have an additional base course of 300 mm depth of lean mix concrete.

Waterbound gravel and hoggin surfaces are for light pedestrian and sometimes for very light vehicular use. These materials are natural in origin: they are dug out of the ground and as such are subject to wide variability.

Drier material may be suitable for light vehicular traffic and hardstanding but excessive use will erode the material and cause rutting. For gravel materials containing a high clay or hoggin content, no vehicular traffic could be possible. One BALI contractor in London once recommended to an architect client (who wished to save some money on a building contract) that Breedon gravel would be suitable for an access lane to a 10 space car park. Unfortunately the material was more than 50 per cent clay and soon became badly rutted and waterlogged. Such intense wear over a narrow drive is not possible with such materials.

Pattern

Pattern is determined by scale and context. Main routes demand directional patterns: brick paving would require 45 degrees herring-bone, instead of basket weave, for example (see Figure 3.57). Large-scale sites require a very simple and bold pattern to be noticed, whereas intimate courtyards have a captive audience and can accommodate fancy, intricate patterns.

Repeating a pattern can provide continuity and rhythm to a design. Patterns require contrast in materials for success: contrast in colour and tone, contrast in type and/or size of materials, bricks against slabs and so on. Patterns may be irregular or regular and symmetrical according to style.

Cost

The cheapest materials are the waterbound natural materials. Concrete laid in situ would be the next cheapest along with the less durable blacktop materials (large areas especially). There are wide fluctuations in price for a particular

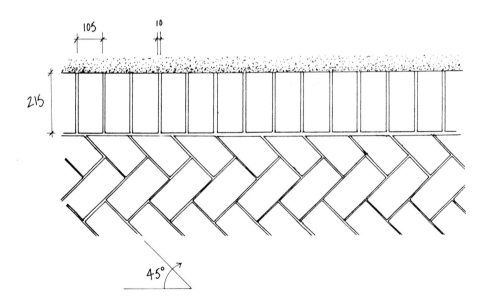

FIGURE 3.57   HERRING-
BONE BRICK PATTERN

paving type, depending upon the quantities involved. Delivery of small loads of paving materials and their base course materials is very much more expensive than large, full lorry loads or complete packs purchased from the manufacturer, rather than from a local builder's merchants. The same is true for different locations. London prices are 25 per cent higher than those of surrounding counties.

Access has a similar effect on price. Restricted access often rules out labour-saving machinery, and narrow pedestrian gates and paths to the rear of terraced property, for example, will mean that materials have to be wheel-barrowed to site, which increases the labour handling time considerably. When you consider that the labour element in any unit price for a paving type is likely to be near to 50 per cent of the total rate (rule of thumb), any factor that increases the labour time will greatly affect the price. Often these factors can combine to create dramatic differences for a specific paving type use on two separate jobs. A difference of £12/m^2 is not uncommon for a simple bitumen macadam surface, one for a large open car park in Suffolk and the other for a narrow path in a London courtyard.

Paving classification

Paving is classified into three broad categories: rigid paving, flexible paving and unit paving. Rigid paving (e.g. in situ concrete) simply means that the paving acts as a solid mass in its response to expansion, contraction and settlement or 'heave'. Flexible paving is the term given to materials that have no tensile strength. Such materials normally have more than one layer, with a base course and a wearing course in addition to sub-base materials. Unit paving is defined as natural stone, baked clay or precast concrete factory produced units, which when laid form a hard surface that is generally impervious.

Rigid paving

Rigid paving materials include in situ concrete, cobbles set into concrete and pointed bricks and setts. If there were any settlement or expansion the material would remain rigid and stable up to the point when the stress was too great and the material cracked. Movement joints are required to reduce the risk of cracking. These are 10 mm joints filled with fibreboard and mastic sealed, which allow the material to expand and contract according to the temperature.

Tamped finish and printed

In situ concrete is a popular material for its cheapness, ease of construction and durability. The strength of concrete used for paving is normally C:20:P mix. In order to ensure that there is no loss of grip in wet weather conditions, the surface is often finished 'tamped' using a board to provide a ribbed or banded texture. The material can be printed and coloured at additional expense to mimic setts, bricks and blocks in various patterns and colours. These printed patterns are usually produced by specialist companies operating as franchises. Although the raw, grey tamped material is harsh and unattractive, the cost effectiveness of

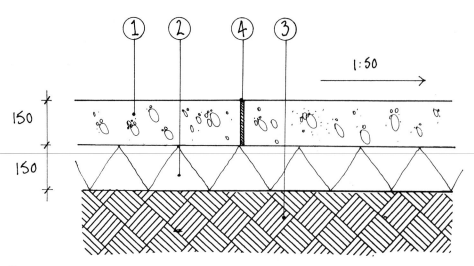

FIGURE 3.58 IN SITU CONCRETE SURFACING

Notes

1. C:20:P mix concrete paving 150 mm thickness with 20 mm maximum aggregate.
2. MOT Type 1 granular fill or similar and approved clean hardcore well consolidated to 150 mm depth to falls and cross falls as detailed.
3. Well-compacted subgrade.
4. 10 mm expansion joints to be incorporated every 5 m.

General notes

- All dimensions in millimetres
- Do not scale from this drawing
- All surfaces to be laid to falls and cross falls as detailed

printing concrete is open to question. When paving large areas greater than 5 m in any direction, movement joints are required, as described above, otherwise the concrete will crack and look unsightly.

Since the material is mostly used for vehicular paving or hardstanding, it is essential to ensure that the depth of concrete is sufficient and that base courses are deep enough and suitably consolidated. For vehicular roads and hardstanding a depth of 200–300 mm is required depending upon traffic load, with a similar depth of well-compacted hardcore or crushed stone over a well-consolidated subgrade. For pedestrian paving the depth of concrete can be 100–150 mm depending upon the amount of traffic and the base course can be the same depth, using hardcore or other granular fill (see Figure 3.58).

Exposed aggregate finish

In situ concrete can be made far more attractive by the use of an exposed aggregate finish. The technique is not easily performed and has to be timed well and carried out by a skilled worker. It is essential to use a gap-graded aggregate, that is, an aggregate of uniform size. This is usually either 10 or 20 mm gauge. The most commonly used aggregate is a washed marine aggregate, as this provides a rich mixture of colours. The concrete is then poured and left to set for 30 minutes. The surface is then washed with a hose and brushed with a stiff bristle brush. This dislodges the cement on the surface to expose the aggregate, which being gap-graded will appear evenly and tightly on the surface. The golden quality of the aggregate will then give the appearance of a gravel surface. The problems with this material are in the damage it can do to bare skin if a child falls over and the lack of skilled people to carry out the work properly. All other factors relating to joints, depths and base courses are the same as above. Further information can obtained from the Concrete Advisory Service division of the Concrete Society.

Cobbles set into concrete

For decorative effects such as edgings, mowing margins or for deterrent paving, cobbles are a useful material, providing a natural stone finish to an inexpensive in situ concrete material (see Figure 3.59). The cobbles are available from marine sources or from riverbeds and are available in a complete range of sizes, 100 mm diameter being the most common. Colours range from the grey to pink of granites from Scottish rivers to the multi-coloured marine aggregates. It is essential to bury 50–70 per cent of the cobble in order to prevent it coming loose in time. The deeper setting will be more appropriate where pedestrian access is necessary, while the prouder fixing will be more suited to deterrent paving. The latter is, in fact, a very effective deterrent to pedestrians, being very hard and uncomfortable to walk on.

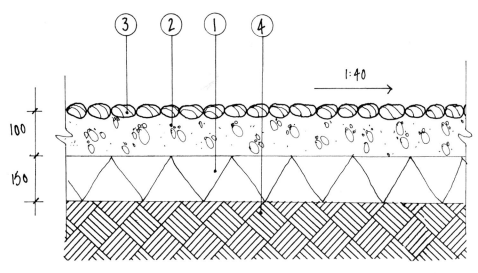

Notes
1. MOT Type 1 granular fill or similar and approved clean hardcore well consolidated to 75–150 mm depth to falls and cross falls as detailed.
2. C:20:P mix concrete bed 100 mm thickness.
3. Scottish beach cobbles laid with axes parallel and flat side uppermost.
4. Well-consolidated subgrade.

General notes
- All dimensions in millimetres
- Do not scale from this drawing

FIGURE 3.59 SURFACE OF COBBLES SET INTO CONCRETE

Flexible paving

Paving is known as flexible when it is able to respond to expansion and contraction with temperature change and also to some degree vertically in the event of differential settlement. Such surfaces include those that have more than one layer, that is they have a base course and a wearing course in addition to any sub-base foundation materials. Such surfaces might be loose fill or incorporate a binder. The binder might be natural (as in the case of hoggin and Breedon gravel) or a manufactured emulsion such as tar or bitumen (as in the case of blacktop materials). Some of the most commonly seen examples of flexible surfaces include bitumen macadam, including coloured versions such as 'red mecamit', and cold rolled asphalt. The strange thing about paving classification is that some unit paving materials are classified as flexible materials, in the same way that pointed brickwork can be described as rigid paving. All dry-lay block paving and dry-lay clay paviors are classified as flexible paving materials because they allow movement without expansion joints. Therefore unit paving is technically a sub-classification of the two main categories of flexible and rigid materials although for reasons of clarity this book sets out paving materials in the three categories.

Blacktop surfacings

Cold asphalt, which is a mixture of bitumen and 6 mm nominal gauge aggregate, is one form of blacktop surfacing. It is used as a wearing course, usually 20 mm thick, laid over a base course bitumen macadam, usually 40 mm thick using a 20 mm nominal gauge aggregate. These materials will be laid over a 150 mm layer of consolidated hardcore or crushed stone all over a consolidated subgrade to cross falls of 1:40. Cold asphalt can be used in quite small areas, unlike hot asphalt, and must be well rolled with a 500 kg roller. It is possible to roll into the surface various decorative aggregates, white, pink or golden as preferred. Though more expensive than bitumen macadam, the material will last at least 10 years.

Coated macadam

Macadam is stone aggregate bound together with a bitumen or tar binder. This material is laid hot and for pedestrian surfaces usually comprises a layer 12 mm thick of 6 mm gauge nominal aggregate wearing course bitumen macadam over a 40 mm thickness of 20 mm gauge nominal aggregate base course bitumen macadam. This is laid over a 150 mm layer of compacted hardcore or crushed stone, all over a consolidated subgrade to cross falls of 1:40 (see Figure 3.60). For vehicular use the thicknesses will be increased to 25 mm, 60 mm and 250 mm respectively (light traffic), with an additional base course of concrete 150–300 mm thick and 150–300 mm thick hardcore (according to the traffic load) for more heavily used roads. For pedestrian surfaces a 500 kg roller is required but for vehicular surfaces an 8000–10 000 kg roller is required.

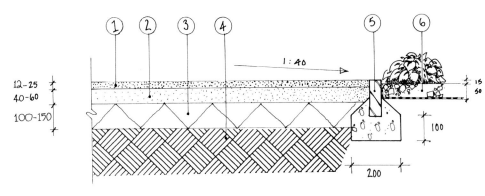

Figure 3.60 Bitumen
MACADAM SURFACE –
PATH

Notes

1. Wearing course of 25 mm depth of 6 mm gauge bitumen macadam, with 3 mm of 3 mm grit to dust rolled in to surface.
2. Base course bitumen macadam, 20 mm gauge aggregate, laid to a consolidated depth of 60 mm.
3. Crushed stone or concrete, hardcore or suitable granular fill-well consolidated to a depth of 100–150 mm.

4. Well-compacted subgrade laid to falls and cross falls as detailed.
5. Precast concrete kerb including strip foundation of C:20:P mix concrete and haunch.
6. Adjacent shrub planting including bark chip and sheet mulch treatment.

General notes
- All dimensions in millimetres
- Do not scale from this drawing

Blacktop with gravel dressing

The above material can be used 'neat' or dressed with a decorative gravel finish (see Figure 3.61). For this material to remain fixed to the bitumen macadam, it must be rolled into the macadam while it is still hot. Otherwise the gravel will sit on the surface and get kicked around on to adjacent grass or planting areas and pedestrians could slip on the loose gravel. In order for the macadam to remain hot, it is essential that the path/road is laid in sections and that the work is carried out during hot sunny summer weather between June and August. The dressing should be 6 mm gauge angular golden gravel and rolled well in with a 500 kg roller (paths). Surplus gravel should be brushed away. The hot rolled asphalt of road construction is finished with larger aggregates, usually 10 mm gauge (but are often 6 mm for private driveways and other roads with light traffic), and these can be the normal grey granite or, better still, golden gravel or red granite to provide a warmer, richer surface.

Coloured macadams

There are many proprietary brands of coloured macadam, which are most commonly either red or green. These surfaces are especially useful where it is necessary to avoid cutting unit paving around corners and where a brighter, more cheerful surface is required than standard blacktop, such as the hard paved areas within play areas.

Hoggin, Breedon gravel and waterbound surfaces

These are materials that are excavated from local pits and, as a result, there is great variability in the nature and colour of the materials. Some will have a

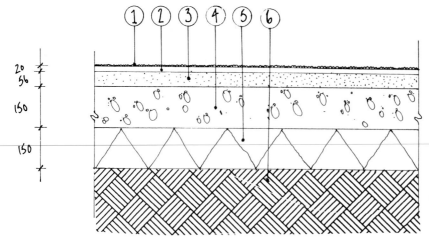

FIGURE 3.61 BITUMEN MACADAM SURFACE – ROAD – LIGHT TRAFFIC. For heavy traffic increase depth of base courses.

Notes
1. Surface dressing of 6 mm nominal size buff coloured gravel chippings bedded in hot bituminous tar.
2. Bitumen macadam wearing course of 6 mm gauge aggregate laid and compacted to a consolidated thickness of 20 mm.
3. Bitumen macadam base course using 20 mm gauge aggregate laid to a consolidated thickness of 56 mm.
4. Lean mix concrete C:7:P mix 150 mm depth.
5. MOT Type 2 granular fill to a consolidated depth of 150 mm.
6. Well-compacted subgrade laid to falls and cross falls as detailed.

General notes
• All dimensions in millimetres
• Do not scale from this drawing

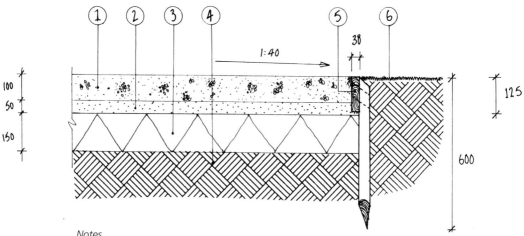

FIGURE 3.62 BREEDON GRAVEL SURFACE – ROAD, VERY LIGHT TRAFFIC ONLY

Notes
1. Breedon gravel wearing course 100 mm thickness.
2. MOT Type 1 granular fill or similar crushed stone well consolidated to a depth of 50 mm.
3. MOT Type 2 granular fill or similar and approved clean hardcore well consolidated to 150 mm depth blinded with limestone dust to falls and cross falls as detailed.
4. Well-consolidated subgrade.
5. 38 mm × 125 mm sawn softwood gravel board attached with galvanized nails to 50 mm diameter × 600 mm long softwood stakes at 1.5 m centres, closer if required at junctions or on curves.
6. Adjacent turf grass surface.

General notes
• All dimensions in millimetres
• Do not scale from this drawing
• All surfaces to be laid to falls and cross falls as detailed
• All timber to be CCA treated to BS 4072 Part 1: 1987 'Tanalized' or similar and approved

higher gravel content than others, while others will have a high clay or hoggin content. It is important that the material is suitably stable for the intended use. Although these surfacings can be used for vehicular hardstanding and very light vehicular traffic, it is essential that the material has a low hoggin content to prevent the surface rutting and becoming a muddy mire. A well-consolidated sub-base of hardcore or crushed stone is essential and it is beneficial also to incorporate a geotextile membrane (see Figure 3.62; please note that this detail excludes geotextile).

These materials are called self-binding, as the application of water will cause the fine hoggin particles to seal the surface. A typical pedestrian specification would be as follows: 20 mm thick gravel wearing course, of material that will pass through a 20 mm mesh, spread and rolled over a 25 mm finished thickness of fine gravel with just sufficient hoggin to act as a binder over a 50 mm finished, rolled, thickness of gravel to pass through a 50 mm mesh, over 150 mm consolidated thickness of hardcore or other approved granular fill. Increase the sub-base to 200 mm for drives and ensure that the hoggin content is low. Roll using a 500 kg roller for pedestrian paths and an 8000 kg roller for driveways.

Crushed stone and stabilized gravel

A durable temporary road or informal driveway can be formed from crushed stone laid in a single layer of 200–250 mm over a geotextile membrane, over a compacted subgrade (see Figure 3.63). This layer should use stone of between 75 to 150 mm gauge. It can be stabilized by a coating of cold bituminous emulsion. A wearing course is then added of 20 mm thickness of 10 mm gauge limestone, granite or other chippings. This layer should be rolled immediately and then rolled again 24 hours later. After two weeks (when it can be open to traffic) the surface should be swept (and sealed with a coat of bituminous emulsion, if required) and coated with a 10 mm thickness of 6–10 mm gauge washed pea shingle.

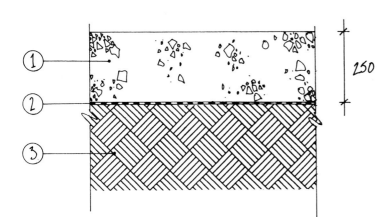

250

Notes
1. Crushed stone surface well compacted to 250 mm thickness.
2. 'Terram' or similar geotextile membrane as blinding to subgrade.
3. Well-compacted subgrade.

General notes
- All dimensions in millimetres
- Do not scale from this drawing

FIGURE 3.63 CRUSHED STONE ROAD SURFACE

The same surface can be provided for informal paths in country parks, conservation areas and gardens. Provide the base as above to 150 mm thickness and spread a 40 mm thick layer of 20 mm gauge chippings, well watered and consolidated. Cover with bituminous emulsion and spread immediately a 12 mm thick layer of 6 mm gauge angular golden gravel. After two days, when the binder has set, a second dressing should be applied and covered with a further layer of the 6 mm gauge sharp gravel.

Crushed brick/cinders and other local materials

Similar informal paths and driveways can be created using other locally available materials, which can be bound or unbound and involve a 100–200 mm thickness of the material well consolidated, laid over a geotextile membrane to blind the 100–200 mm thick sub-base of crushed stone or hardcore and also to prevent weed growth in the path (to blind is the term used to describe the filling up of spaces in between aggregates used for base course materials).

Unit paving

There are nine main types of unit paving, each of which must be laid over an appropriate sub-base course material, the thickness and type depending upon whether it is vehicular or pedestrian paving.

1.   Brick paving.
2.   Concrete blocks and clay blocks or paviors.
3.   Concrete sett paving.
4.   Precast concrete slab paving.
5.   Concrete fire-path units.
6.   York stone flag paving.
7.   Stone sett and block paving.
8.   Tiles.
9.   Wooden sett paving.

Brick paving

A brick is a clay unit 215 × 105.5 × 65 mm and is the same as a house brick, except that paving bricks have to be very hard to ensure that they do not get frost damaged. Bricks do not have the bevelled edge of dry-lay paviors and so they must be pointed up using a cement mortar, as rigid paving. This is a fiddly and time-consuming job, which can all too easily be carried out badly, causing cement stains to the brick surface. Bricks themselves are expensive too, especially the more attractive stocks, so this type of paving is certainly a luxury. Having said that, well-pointed brickwork using an attractive multi-stock brick is one of the most attractive of any paving materials, in both traditional and modern contexts (see Figure 3.64). Bricks should be bedded on either a sharp sand bed (50mm) over 100 mm of hardcore for pedestrian traffic, or sand/cement bed 1:5 (50 mm) depth over 250 mm or more for light vehicular traffic.

Heavier traffic would require a concrete base above the hardcore of 150 mm or more for a busier road. If the bricks specified are too soft, then water will penetrate in the winter time, freeze and break up the surface of the brick.

Concrete block paving and clay block paviors

A block is a unit 200 × 100 × 65 mm and the most commonly used blocks are made from pressed concrete, though the much dearer clay blocks are more attractive. All blocks differ from bricks in several ways. Apart from the obvious difference, being made from concrete instead of clay, blocks are slightly smaller than bricks. Clay blocks are sometimes called 'clay paviors' but are just the same material.

The concrete blocks can be coloured with artificial dye and they are available in natural grey, red, buff, black, green, brown and 'brindle', which is a mix of red and black to mimic (badly!) the effect of stock

FIGURE 3.64 PAVING CONSTRUCTED FROM POINTED STOCK BRICKS

bricks. These colours leach out and fade quickly, especially in areas of high wear. Clay blocks have an intrinsic colour of the burnt clay and therefore never fade. Some have interesting mixed colours, of red and blue like those of some stock bricks, but with the benefit of being much harder wearing. However, the dry-lay blocks are much harsher, modern and more urban in appearance than the mortar-pointed stock brick (see Figure 3.65).

All blocks have a bevelled edge and spacing ridges along the sides so they must be laid dry and never pointed. Sand is vibrated into the joints to bind them. They are therefore easy and quick to lay, and concrete blocks have a relatively low cost (compared to bricks) which makes for a highly durable and cost effective surfacing, both for pedestrian and vehicular surfacings (see Figure 3.66). However, there is no substitute for clay blocks and bricks in aesthetic terms, though the price is high. Blocks are bedded on either a sharp sand bed (50 mm) over 100–150 mm of hardcore for pedestrian traffic, or sand cement bed 1 : 5 (50 mm) depth over 250 mm or more for light vehicular traffic. Heavier traffic would require a concrete base above the hardcore of 150 mm, or more for a busier road. Blocks are very hard, even the clay ones, and absorb

FIGURE 3.65   DRY-LAY
CLAY PAVIORS

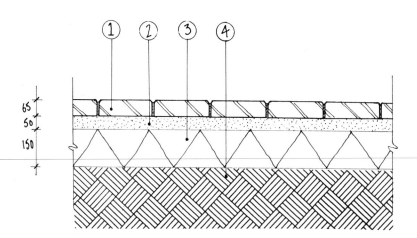

Notes

1. Ockley dry-lay paving units 200 × 100 × 65 mm laid brick on flat, butt jointed with kiln dried soft sand brushed into the joints.
2. 50 mm thickness consolidated sharp sand.
3. MOT Type 1 granular fill or similar and approved clean hardcore well consolidated to 150 mm depth to falls and cross falls as detailed.
4. Well-compacted subgrade.

General notes

- All dimensions in millimetres
- Do not scale from this drawing
- All surfaces to be laid to falls and cross falls as detailed

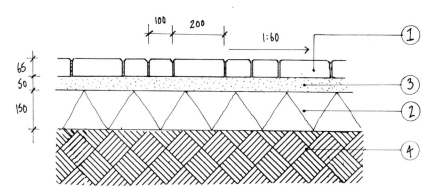

Notes

1. Precast concrete blocks 200 × 100 × 65 mm laid in a herring-bone pattern and butt jointed with kiln dried sand brushed into the joints.
2. MOT Type 1 granular fill or similar and approved clean hardcore well consolidated to 150 mm depth to falls and cross falls as detailed.
3. 50 mm thickness consolidated sharp sand.
4. Well-consolidated subgrade.

General notes

- All dimensions in millimetres
- Do not scale from this drawing
- All surfaces to be laid to falls and cross falls as detailed

FIGURE 3.66   DRY-LAY
PRECAST CONCRETE BLOCK
PAVING

little water, so there is no risk from frost or stiletto heels, which can damage softer stock bricks.

Concrete sett paving

A concrete sett is a variable sized unit but most are 100 × 100 × 50 mm and are often shaped to allow curved patterns, fan patterns and so on. They are dearer than concrete blocks, though a little cheaper than clay blocks and bricks. The concrete setts can be coloured with artificial dye and they are available in natural grey, pink, buff, black, brown and green. Like concrete blocks, these colours leach out and fade quickly, especially in areas of high wear.

The setts are designed to mimic natural stone setts but their uniformity fails to achieve this. Notwithstanding this fact, they have their own aesthetic, which is attractive for urban pedestrian paving, where budgets are fairly generous. They are perhaps surprisingly less suited to traditional areas than to more modern contexts. Concrete setts are dry-laid with sand vibrated in to bind them. They are bedded on either a sharp sand bed (50 mm) over 100 mm of hardcore or directly on to a screed bed of concrete (150 mm depth pedestrian and light vehicular traffic, 250 mm for heavier vehicular use, when a further 250 mm of hardcore sub-base would be needed).

Precast concrete slab paving

There are so many shapes and sizes of precast paving slab that you can choose one to match most contextual settings, from rustic and rural to high tech. The most common slab is the standard pressed 600 × 900 × 50 mm or the 600 × 600 × 50 mm grey slab. These represent a cheap and easy to lay paving surface that can be laid dry and butt jointed, with sand/cement brushed into the joints, or pointed up with cement mortar with a 10 mm wide joint as preferred.

For pedestrian purposes the slabs can be bedded on 50 mm of sharp sand, sand/cement mix or screed, over 100 mm of hardcore. Such slabs can be coloured using artificial dyes and are available in all the colours mentioned for concrete setts. Where vehicular traffic may occasionally drive over the slabs, authorized or not, it is essential to use a 70 mm thick slab. In such situations the sub-base would also have to be deepened to 250 mm with possibly an additional layer of concrete, if this vehicular traffic was regular and of heavy load.

Slabs are commonly available in a variety of sizes additional to those mentioned above. A particularly common size for paths and patios is the 450 × 450 × 50 mm slab but some are available in 300 × 300 × 50 mm (or 70 mm thick). Marshalls produces a slab with a York stone aggregate (called 'Perfecta') which is more attractive than the pressed grey slabs and can be used in most contexts. It is available in natural grey, buff, red and green. A rough, tooled surfaced slab is also available ('Saxon', also from Marshalls) that is particularly good for its non-slip properties. Other types of more ornamental slab, some using stone aggregates, are designed to mimic natural stone (with varying degrees of partial success), possessing a 'riven' irregular surface supposedly like the grain of natural stone.

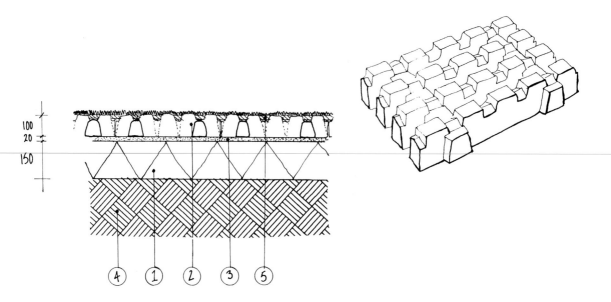

Notes

1. MOT Type 1 granular fill or similar and approved clean hardcore well consolidated to 100–150 mm depth to falls and cross falls as detailed.
2. 450 mm × 450 mm × 100 mm precast concrete fire-path units interlocked in accordance with manufacturer's instructions.
3. 20 mm thickness consolidated sharp sand.
4. Well-compacted subgrade.

5. Topsoil/peat mix infill to be spread to fill pits completely, filling and covering fire-path units, seeded with hardwearing amenity grass seed mix.

General notes
• All dimensions in millimetres
• Do not scale from this drawing

FIGURE 3.67 FIRE-PATH UNIT PAVING

Concrete fire-path units

Fire-path units are rectangular units that have a rigid structure, with gaps or holes to roughly 50 per cent of the surface area, which can be filled with soil and seeded with grass (see Figure 3.67). They provide a grass effect that can be driven over by fire engines or maintenance access vehicles. They are often used for rural car parks. The units are bedded on a 150 mm consolidated layer of hardcore or crushed stone, over a consolidated subgrade. The grass tends to die back severely if wear is too severe and this material has been superseded by proprietary rubber stabilization materials which are 90 per cent soil and grass, rather than 50 per cent, and as such are less prone to grass die back. Indeed, there are materials available now, such as Fibresand, that allow the weight of a fire engine over a seeded root zone material containing a high sand content, mixed with artificial fibres.

York stone flag paving

Flagstones obtainable from quarries show a considerable variability in size as they are a natural material. The thickness of such flags can vary between 35 and 65 mm. York stone flags can be bedded as for precast concrete slabs (see Figure 3.68).

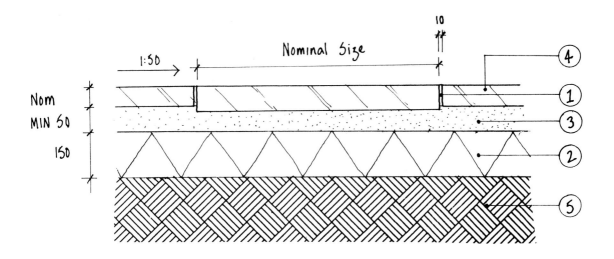

Notes
1. 1:3 mix cement mortar pointing (and haunching).
2. MOT Type 1 granular fill or similar and approved clean hardcore well consolidated to 75–150 mm depth to falls and cross falls as detailed.
3. Minimum thickness 50 mm consolidated sharp sand bed.
4. Nominal size York stone flags.
5. Well-consolidated subgrade.

General notes
• All dimensions in millimetres
• Do not scale from this drawing
• All surfaces to be laid to falls and cross falls as detailed

When using reclaimed flagstones there is even less control over the size of the units. Despite these factors, there is no better material for traditional streets, formal squares and pond surrounds and it is at home in historic towns, villages and most gardens. Flagstones can look very out of place in modern cities and I recall a large number once used at a bus stop on the Isle of Dogs, East London, which looked positively incongruous and therefore unnecessarily expensive.

FIGURE 3.68 YORK STONE FLAG PAVING

Stone sett and block paving

Stone setts are far more commonly seen in the rest of Europe than in the British Isles, where paving tends to be fussier, utilizing generally smaller units.

It is a common feature of many German squares to see elaborate fan patterns using different colours of granite sett, often 100 mm cubes or even smaller. Setts, cobbles and stone blocks are seen most frequently in towns in northern Britain such as Halifax, where the natural material (millstone grit and granite) is readily available. Stone setts are normally 100 mm cubes and are available in new and reclaimed forms. They are often used as surrounds to trees, lighting columns and manhole covers, being suited to tight radii. They provide a slightly uneven but non-slip surface that is very attractive in traditional urban settings such as historic towns, villages and some gardens.

FIGURE 3.69 POINTED
SETT PAVING

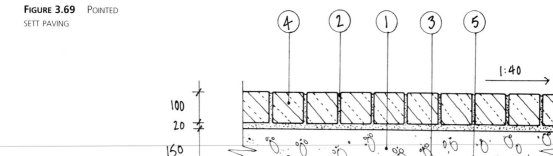

Notes

1. Concrete foundation of 150 mm thickness C:20:P mix.
2. 1:3 mix cement mortar pointing, bedding and haunching.
3. MOT Type 1 granular fill or similar and approved clean hardcore well consolidated to 75–100 mm depth to falls and cross falls as detailed.
4. 100 mm × 100 mm × 100 mm granite setts, bedded, pointed and flush pointed on 1:3 mix cement mortar.
5. Well-compacted subgrade.

General notes

- All dimensions in millimetres
- Do not scale from this drawing
- All surfaces to be laid to falls and cross falls as detailed

Though they can be laid dry with sharp sand brushed into the joints, stone setts are often pointed and jointed with cement mortar, 1:3 mix (see Figure 3.69). Laid dry, they are useful as tree surrounds in urban areas when bedded on 6 mm pea shingle to allow water percolation to the tree pit. The material is less susceptible to vandalism if mortar jointed, however, as once one sett has been removed (with dry-lay situations) the entire paved area can be easily lifted, used as missiles or stolen.

Once the staple paving material for city streets (before the use of tarmacadam), stone blocks, 200 × 100 × 100 mm, often of granite or millstone grit, provided quite a ribbed and uneven surface which was hard to walk on, especially with narrow heels. Often these stones became rutted by the coach and horse transport of the day. These blocks are most often available secondhand from reclaim companies but can be purchased new. They are suited to restoration or conservation use in historic towns, cities and villages.

Tiles

Exterior grade tiles are made from clay and can be a variety of sizes and colours. Comparatively few are frost resistant and great care should be taken to ensure that they are suitable. A further problem with some tiles is that they

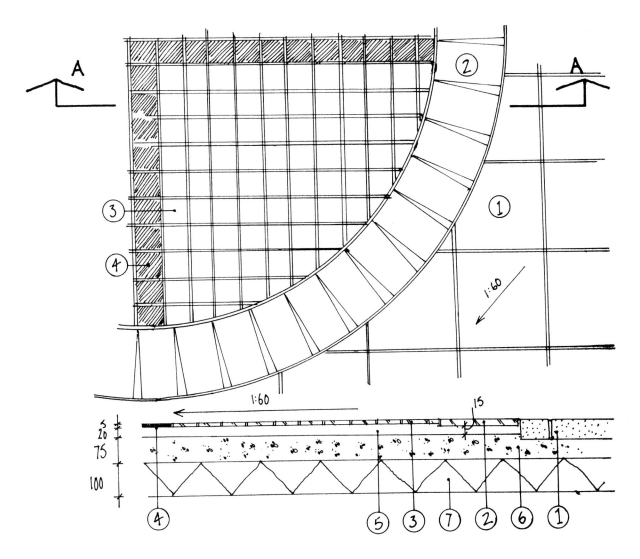

Notes

1. Surrounding paving/concrete block paving surface.
2. 240 mm 'Cyclamen Red' 1100/060 tiles, 10 mm thick by National Ceramics Ltd. Tel (01552) 625162.
3. National Ceramics 'Jasba' white non-slip tiles 75 mm x 75 mm.
4. National Ceramics 'Jasba' blue (floor tile non-slip) 4423.
5. Arduit X7 tile adhesive 20 mm thickness (15 mm with 'Cyclamen Red') to manufacturer's instructions.
6. Sand/lime/cement semi-dry mix, using sharp sand – zone 2 to 3.
7. Suitable crushed stone base material well consolidated to falls and cross falls.

General notes

- All dimensions in millimetres
- Do not scale from this drawing
- All tiles to be treated with 'Fleck Stop' impregnating dressing, to manufacturer's instructions
- Mix grout and Arduit C2-grey with added Ardion 101 to manufacturer's instructions

FIGURE 3.70
DECORATIVE CERAMIC TILES

become slippery when wet. This is less of a problem with quarry tiles, which have a rough surface, but some of the more ornamental, coloured tiles may cause problems. These brightly coloured tiles, available in white, blue, red and green, can still be used in bands within more common paving materials such as slabs and blocks, where they provide a striking accent of colour to what would otherwise be a more ordinary paved area (see Figure 3.70). Tiles should be fixed to an even concrete screed with a proprietary cement-based tile adhesive.

Joints should be 6 mm wide and grouted with a proprietary grouting material, coloured to suit the tiles. The tiles should then be sealed with a proprietary sealant to prevent discoloration. All these processes ensure that tiles can be an expensive alternative to choose but for the urban, modern and high tech situation they can be outstanding and are under-used.

Wooden sett paving

Wood is an unusual material for paving as it is inclined to rot and becomes very slippery when wet. However, it is possible to obtain treated wooden setts which, being small individual units, provide more grip. Additional grip can also be provided by stapling chicken mesh to the surface, as used for timber decking and bridges. If this is used, it is important to use galvanized mesh and staples. Wooden blocks are a material mostly confined to garden usage.

Timber setts should be bedded on sharp sand over a compacted subgrade, and a geotextile membrane is advisable, such as ICI 'Terram 1000' or 'Mypex' sheeting, to be laid over the subgrade. In fact, the use of geotextile fabrics should be considered for all the above unit paving surfaces to prevent weed growth, without the need for hazardous residual herbicides.

Drainage

This section could be the subject of a book in its own right and as the available space is not sufficient to cover the subject in great detail, the following selective passages merely offer useful information for everyday landscape schemes.

Surface water drainage

Surface water drainage is the most commonly encountered aspect of drainage in landscape schemes. Very often insufficient attention is paid to the shedding of rain-water from paved surfaces and the result is semi-permanent, muddy puddles, which make the paved surface slimy, grimy and sometimes impassable. When surface water drainage is poor, paving slabs may develop hollows where the base has been washed away and the slab soon begins to wobble underfoot. With each pedestrian passing, more base material is displaced and more water let in. Eventually, every time a pedestrian treads on the slab, a spurt of water will shoot up their legs. There can be few of us who have escaped this experience first hand.

Falls and cambers

In order to shed water from paths, pavements or roads, a slope or fall is required. These falls come in two types: longitudinal falls and cross falls (see Figure 3.71).

Longitudinal falls run along the full length of the pavement and can be quite shallow, from 1:50 near buildings and smaller paved areas to 1:250 for drainage channels, gutters and pipes.

Cross falls run at 90 degrees to longitudinal falls and may be 1:30 for rough surfaces, 1:40 for most paths, cycle tracks and roads, but for large areas of pressed concrete slabs, the minimum cross fall could be as shallow as 1:60–1:70. Rougher surfaces, such as waterbound gravel, hoggin, etc., stock brick and bitumen macadam, require cross falls from 1:30 (gravel) to 1:40 for brick and bit'mac'. For engineering bricks and paviors and for in situ concrete, 1:60 would be the minimum fall recommended. For roads and wider paths it is not practical or desirable to have a single cross fall from one side to the other so a camber is constructed, with a high point in the centre of the road, shedding water to both sides. With large paved areas, too, the designer will need to think

FIGURE 3.71   LEVELS AND FALLS TO DRAINS

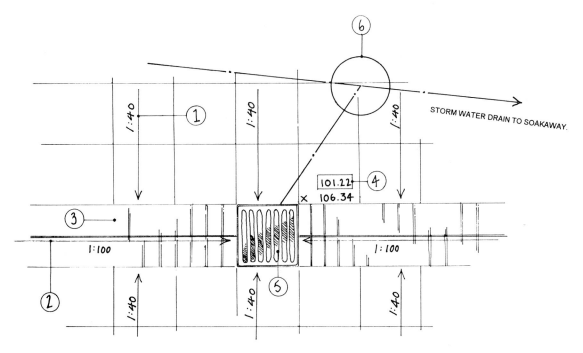

Notes
1. Cross falls of 1:40 on paved surface to longitudinal falls.
2. Longitudinal falls 1:100 along channels to gulleys.
3. Bricks laid flush pointed to form channel.
4. Existing and proposed spot levels: existing in box, proposed unboxed.
5. Grille to Deans gulley leading to storm water run-off main drain.
6. Manhole cover to storm water drain leading to soakaway.

General notes
- All dimensions in millimetres
- Do not scale from this drawing

about high and low points and ensure that the cross falls shed to central channels, just like a series of pitched roofs falling to central gutters.

Gulleys, channels and gutters

To determine what happens to water when it is shed from the main paved surface is just as important a consideration as the provision of suitable falls in the first place. The key factor here is the edging treatment and the size of the paved surface area. Clearly, vast paved surfaces, such as those of Trafalgar Square or Moscow's Red Square, produce incredible volumes of surface water in times of heavy rain. It is essential for the drainage to be able to cope with a 1 in 50 year storm. With large paved areas (even large terraces, wide streets, play areas and squares) there will be too much water simply to shed the water to grass or planted areas without causing a flood and/or waterlogging the planting beds or grass areas.

FIGURE 3.72 GRANITE SETTS USED AS SURFACE WATER RUN-OFF CHANNEL IN PEDESTRIAN PAVED AREA

Channels and gutters will be necessary to collect the surface water running off the paving. These will be constructed from brick, granite kerbs or setts recessed or dished and set on an in situ concrete foundation (see Figure 3.72). These channels will fall longitudinally, usually at around 1 : 100 or else to the lie

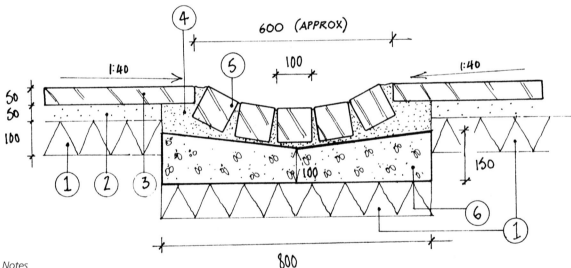

Notes

1. MOT Type 1 granular fill or similar and approved clean hardcore, well consolidated to 75–100 mm depth to falls and cross falls as detailed.
2. 50 mm thickness sharp sand.
3. 450 mm × 450 mm × 50 mm precast concrete slabs laid jointed and flush pointed with 1:3 mix cement mortar.
4. 1:3 mix cement mortar. Joints shall average 20 mm (maximum width 30 mm).
5. 100 mm × 100 mm × 100 mm granite setts, bedded, pointed and flush pointed on 1:3 mix cement mortar to form rain-water gulley. All laid to give longitudinal falls of 1:100.

NB Narrower channel is commonly used with 4 setts instead of 5.

6. 150 mm × 800 mm strip foundation of C:20:P mix concrete.

NB This detail is also suitable for use adjacent to road and drive surfaces. It can be constructed using 4 setts.

General notes
- All dimensions in millimetres
- Do not scale from this drawing
- All surfaces to be laid to falls and cross falls as detailed
- Joints to be not more than 30 mm

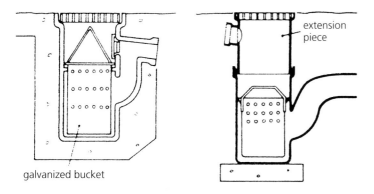
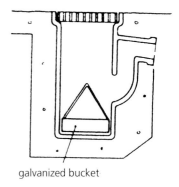

galvanized bucket   extension piece   galvanized bucket

of the land. At intervals (usually 45 m with roads) gulleys will be set into the channel to take the running water away to a surface water drainpipe. This surface water may well join the foul water drainage system in some instances.

FIGURE 3.73   DEANS GULLEY

Certainly, any large area of paving (say greater than 5 m × 5 m) will need gulleys, as will any paving that has an edge treatment that is proud of the pavement surface level. All kerbs, brick or concrete edging units or timber edgings that are anything but entirely flush with the main paving surface will obstruct the passage of water to adjacent soft surfaces and will therefore need gulleys at intervals along the kerb or edging to take the water away.

A gulley is a lined pit that has a metal frame and grating which allows water to fall into it. A pipe leading from the gulley takes the water away to a soakaway, existing drainage system or water course (see Figure 3.73). At the point where the water leaves the pit, the pipe will have an 'S' bend in it to provide a water trap, like that of a toilet, to prevent gases and odours rising from the gulley. This pipe will not be at the bottom of the gulley pit but at least 300 mm up the side. This position allows all suspended solids in the surface water to settle out at the bottom of the gulley so that they do not block the pipes. Some gulleys have a bucket at the bottom to collect the 'gunge' and allow its easy removal and disposal, otherwise it has to be sucked out by a vehicular roadside vacuum cleaner. The lining to gulleys can be metal or engineering brick or concrete.

The metal frame is fitted (in the case of proprietary metal gulleys) or set in cement mortar (1:3 mix) over brick or concrete linings. For square or rectangular gratings or, indeed, inspection covers, the frames should be set in line with the direction of the pavement, both for aesthetic reasons and to avoid cutting of unit paving materials around the grating or cover.

Inspection covers

Generally inspection covers should be set flush with the paving surface and should be sufficiently heavy duty to take the type and volume of traffic over them. Where funds allow, inspection covers can be recessed to take a proprietary metal frame, which fits over the inspection cover and takes paving materials to match those of the main paving, preventing such covers despoiling the look of the paving. You can, of course, actually make a feature of inspection covers and gratings by designing unit paving surrounds to them. Gratings, of course, need to be recessed just slightly (15–25 mm) to receive the water flowing down channels or along edgings.

Pipe sizes, junctions and changes of level

Surface water drainage pipes in pedestrian pavements leading from small gulleys will usually be 100 mm diameter. Where these connect with one or two other such pipes, the diameter will increase to 150 mm. Then when two such larger pipes join, the diameter will increase to 200 mm and so on. The design of the piping must be carried out according to the individual requirements of the site, however. Pipe junctions and any changes in direction will require an inspection or rodding chamber to enable blockages to be cleared. Typical falls will be 1:150–1:250. Such pipes should be 600 mm below ground to prevent heave and damage in frosty winter weather, though heave after significant rainfall following a prolonged drought in clay soils may cause less avoidable damage.

Avoiding damp

Whatever type of drainage the designer chooses, it is important to ensure that, wherever possible, surface water drainage falls (that is, the paving gradients) are away from buildings to avoid any risk of damp. Such falls near buildings should be a minimum of 1:50. Paving surfaces should be 150 mm below damp proof courses of buildings to prevent any risk of rising damp.

Council approval

All paving schemes involving surface water drainage pipes and gulleys need approval from the local council and where such schemes involve existing water courses the Environment Agency will need to approve them.

Cost implications

The problem with the use of gulleys and drainage pipes is that they are extremely expensive and such drainage provision can double the cost of paving a space. Clearly, where the areas of paving are not too large and where budgets are tight, it is often adequate and/or necessary to construct the path or paving with cross falls (1:40 typically) to the soft landscape areas.

Falling paving to soft surfaces

If it is decided that surface water will be shed to the planting or grass areas, then there are many matters to consider before details can be prepared. It is essential that any edgings proposed to the paved surfaces are flush with the paving and that all planting bed soil levels are 65 mm below pavement levels to receive the water and to allow for sheet and/or 50 mm depth of bark mulch to be laid over, and still leave a recess of at least 15 mm. This will ensure not only that the water is received from the paving but also that the bark does not blow around over the pavements and cause problems with maintenance.

With adjacent grass surfaces, contrary to all conventional detailing (which is flawed), it is essential that the soil levels (of either turf or seeding) are 5 mm (but no more) below adjacent paving. This is sufficient to receive the surface water but not enough to cause mowing problems. Convention dictates that the soil levels are 25 mm higher than the pavement, in order that the lawn edges can be cut neatly by placing the wheel of the mower on the paving. This is only acceptable, however, if there are surface water drainage gulleys, otherwise the raised soil will act like a dam and puddles will form.

If the surface water is to be successfully shed to the lawn areas, then the lawn soil level must be lower than the paving. The danger is that if the levels are too low, anything below 10 mm, the mower blades will either hit the pavement edge or leave a tuft of longer grass along the edges which will look unsightly.

The fact remains that with the majority of landscape schemes budgets do not include sufficient money for drainage gulleys and pipes and in these situations very careful attention to the levels and the specification of finished levels is essential to avoid ponding, puddles and stained, slimy and therefore slippery paving surfaces.

If the surface water is to shed to the beds and the levels are addressed accordingly, there is still a potentially serious problem to resolve. The designer must be fully aware of the nature of the soil throughout the site. Heavy, sticky clay soils are not suitable for receiving large amounts of surface water. In the winter months (at least) the beds or lawn areas will become waterlogged and plant roots will rot and the plants die. Grass areas will become moss bound and the grass thin. With any wear, the grass areas will easily compact and the grass will soon fail in the wet, compacted soil.

Even soils that do not contain a high clay content can be compacted by building operations and, as often happens, the topsoil may be spread over the compacted subgrade, causing the planting bed to act like a big bucket and fill up with water. With soils other than clay, as long as there is no such compaction (or as long as the main contractor has broken through compacted ground before topsoiling), the bed will drain freely and surface water can satisfactorily be shed on to the beds. It is worth noting that it is by far the best policy to ensure that the landscape contractor, wherever possible, is responsible for topsoil provision, both to ensure that the subgrade can be inspected for contamination and compaction following building activities and also to prevent main contractors from burying rubble and rubbish in the planting beds.

Surface water drainage and trees

Surface water drainage for new paved areas can have a significant effect on trees in development schemes. The water table inevitably falls as the surface area available to absorb rainfall decreases. Trees will be put under stress until their root systems can adapt and find the water at a deeper level. Older trees (which are often the most attractive and have the greatest landscape impact) are often less able to adapt to such changes and may well die. There is therefore an increasing trend for drains to lead to soakaways or French drains so that the water can be absorbed into the ground water and maintain the height of the water table.

Soakaways and French drains

A soakaway is a large pit (size to be determined by the area to be drained) filled with rounded clean aggregate (see Figure 3.74). The base of the pit should be of a free-draining material. The sides, base and top of the pit are best wrapped in a geotextile woven fabric to prevent soil clogging the spaces between the aggregate. Most soakaway pits will be between 1 and 3 m cubed but are sometimes larger.

A French drain is a useful drainage method for larger paved areas, especially in rural locations or adjacent to car parks, at the base of embankments, etc. French drains are essentially ditches 300–450 mm wide with sloping

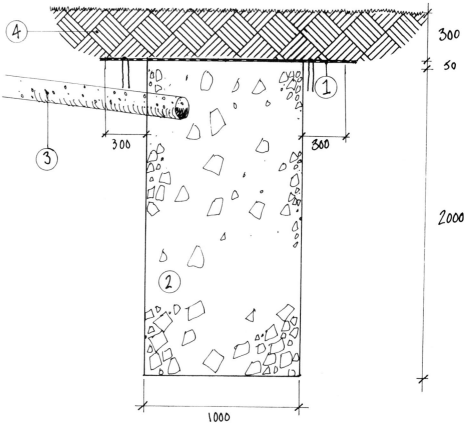

FIGURE 3.74
SOAKAWAY

Notes

1. 'Terram 1000' geotextile membrane pegged with galvanized metal pins at 500 mm centres.
2. 2000 × 1000 × 1000 mm chamber infilled with washed 20–50 mm gauge shingle or marine cobble rejects.
3. 110 mm drainage pipe from storm water run-off.
4. Topsoil backfilled over top of soakaway cap and re-seeded or planted to blend with adjacent surface.

General notes
- All dimensions in millimetres
- Do not scale from this drawing

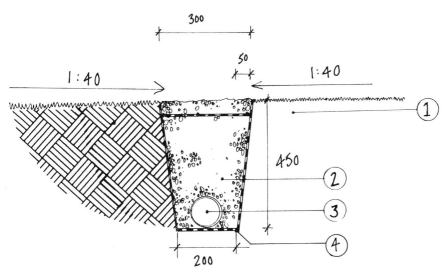

FIGURE 3.75 FRENCH DRAIN

Notes

1. Adjacent topsoil surface falling to drain channel.
2. Washed pea shingle infill to French drain channel. Pit sides dug at angle to prevent slippage (according to angle of repose).
3. 110 mm diameter perforated plastic drainage pipe laid to longitudinal fall of 1:150 minimum.

4. Geotextile fabric – 'Terram 1000' or equal and approved.

General notes
• All dimensions in millimetres
• Do not scale from this drawing

sides, filled with clean rounded aggregate (see Figure 3.75). The bottom, sides and top of the ditch will be wrapped in a woven geotextile fabric and a 75 mm decorative top of pea shingle will be added to conceal the fabric.

An 80–110 mm diameter plastic perforated land drainage pipe can be added at the bottom of the French drain to help spread the water along the ditch or to take excess water to a water course. Mostly, however, the water entering the French drain seeps away along the length of the ditch and so the ditch acts like a long soakaway. The ditch side will be sloped at a gradient governed by the angle of repose for the soil in question. The sides of the drainage trench can be stabilized by using geotextile fabric material over the face of the slope and it is often used to line the base of the drainage trench and to blind the pea shingle fill material.

Foul water drainage

Foul water drainage is the drainage of sewage effluent from buildings to the main drainage system and then to a sewage works or, where main drainage is not available, the foul water effluent is drained into a septic tank or cesspool. A septic tank is a large container that receives the raw sewage. The solids settle out and are broken down by anaerobic bacteria, while the fluid is allowed to soak away into the ground water or is channelled to a coke filter bed, where aerobic bacteria digest the foul elements. From time to time the solids have to be removed, otherwise the tank will fill up.

Pipe sizes, inspection chambers, gulleys and connections with foul water drainage are all governed by the local authority building regulations and those of the water authority, and it is rare that a landscape designer will be involved with this area.

Land drainage

Land drainage can be split into three categories: agricultural, general landscape and amenity, and sports turf drainage.

Agricultural

Agricultural drainage is the cheapest form of land drainage, involving a tractor-mounted mole drain tool (which is a rocket shaped metal device attached to the base of a blade), which cuts through the soil to a depth of 600 mm and leaves a round hole in the soil. This is carried out at between 1 and 3 m centres, depending upon wetness and funds available. The water flowing down the mole drains must then discharge into a ditch or piped drain. Mole drainage lasts between three and five years, before the holes fill in and the process must be repeated.

Landscape and amenity

For landscape and general amenity areas or for wetter agricultural areas and where the budget is more generous, plastic perforated land drainage pipes are used, normally laid out to a herring-bone pattern. These 80 mm diameter drains connect with a central main drain.

Trenches 150 mm wide are dug by a trenching machine which also lays the pipe. The trench is then backfilled with crushed stone aggregate and usually topped with sand. These drains are spaced at between 3 and 10 m intervals, according to the wetness of the land and the budget available. Land drains will be laid at between 450 and 600 mm depth with falls of 1:250, or else to the lie of the land. All the land drains will connect with the central main drain of 110–225 mm diameter (the larger diameter drains being at the end of a run, to take the progressively increased flow), which will discharge into a water course, ditch or surface water drainage system. The trenches can be lined with a geotextile fabric to lengthen their working life. Land drainage should last many years, especially when geotextile fabrics are used to prevent soil clogging the pipes and aggregates.

Sports fields

For sports turf areas a grid pattern of drains may be used so that the main drains can be positioned on the outside of the playing fields (see Figure 3.76). Where finer grade surfaces are required throughout the year and where more generous budgets are available, sand slit drains can be added. Sand slitting is the term for 250 mm deep and 50 mm wide slits filled 175 mm with washed,

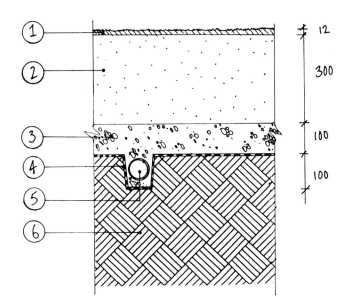

Notes
1. Upper horizon of 'washed turf'.
2. USGA specification rootzone material 70/30 or 80/20 sand/soil mix.
3. 6–10 mm nominal aggregate washed and rounded pea shingle as drainage layer.
4. Geotextile fabric – 'Terram 1000' or similar.
5. Plastic perforated drainage pipe 100 mm diameter.
6. Well-compacted subgrade.

General notes
- All dimensions in millimetres
- Do not scale from this drawing

rounded granular aggregate and 75 mm of sharp sand. These slits are spaced at 1 m centres and are prepared at 90 degrees to the deeper land drainage trenches; the stone fill of the sand slitting allows the excess water to drain to the main land drainage system. Sand banding is a similar method but the bands are 20 mm wide and spaced at 500 mm centres. The soil profiles themselves can be given drainage layers, more especially for the higher grade pitches. The design and construction of such sports fields is very much a specialist area. When contemplating the design of sports areas it is worth commissioning advice from the Sports Turf Research Institute.

FIGURE 3.76 USGA SPORTS FIELD SOIL PROFILE

Other features

There are many further features of hard landscape design such as lighting or special constructions, gazebos, arbours, pergolas, ponds and so on that cannot be covered in any detail in an introductory book of this type. These items are either purchased as 'off-the-peg' proprietary units or are to some extent designed individually.

Lighting

Lighting is a specialist and massive subject and though I feel that lighting requires greater attention from landscape designers, there are specialist publications that can be consulted. Some of the proprietary lighting companies can also be of great assistance in preparing workable lighting schemes.

It is often only in private developments and gardens that lighting design can be taken to its full potential. Any public highway will be adopted by the local authority who will have restrictive policies about the lighting units because of maintenance and cost. As such many public lighting schemes are destined to have a somewhat municipal appearance. A typical amenity lighting scheme likely to be approved by a local council might include a 5 m high steel column from Abacus or a similar company spaced at 30 m intervals and fitted with a CU Phosco 'P107/24 Severn' lantern with a 70 watt SON bulb. This should provide the minimum 1 lux between columns and prevent 'light pooling', where an over-bright bulb causes the eye to see the spaces between the lanterns as very dark compared to the light areas, and therefore threatening. The advantage of these light fittings is their low cost and their top cover. This cover ensures that the lighting scheme should conform to 'Dark Skies' policies.

Obviously, on private land the designer is free to choose and design much more elaborate and expensive lighting schemes. Such schemes might include up-lighting or down-lighting the façades of buildings or indeed trees, which can make an impressive display using external grade floodlights. Bollard lighting units can be employed to punctuate a pedestrian path like a series of hovering fireflies.

Gazebos and summerhouses

These architectural features can be purchased from specialist companies (such as Ollerton) or designed specially. They can be constructed from either metal or wood and can be quite elaborate affairs, with power, water and drainage provided. They are always an asset to the larger garden, providing interest and a useful focal point.

Pergolas and arbours

Pergolas are invariably timber constructions on which to support climbing plants (see Figure 3.77). They stand approximately 2.5 m tall with the lowest cross braces at around 2.1 m. The simplest constructions are invariably the best, though there are some elaborate designs produced by specialist companies. Timber always warps to some degree when fully exposed to the elements, and the simple provision of two 50 mm × 175 mm rails twice bolted either side of a 125 mm square post using galvanized coach bolts is entirely adequate and very sturdy (see Figure 3.78). The pergola posts and rails usually stand each side of a path or patio and are braced with cross rails of similar dimensions to the longitudinal rails, positioned above or below these rails.

All the rails and braces should oversail the posts by at least 600 mm and the appearance of the structure can be improved by providing a chamfered edge to these protruding rails, as well as post caps or finials to the tops of the posts. All timber will normally be pressure-treated sawn softwood, which can be stained with a dark preservative wood stain of a non-phytotoxic kind (such as Sadolin 'Classic'). Posts are best set into metal shoes above ground and these shoes set into a concrete pad below ground.

If climbing plants are to twine over the pergola, plastic-coated, heavy

FIGURE 3.77 PERGOLA.
This basic design can
be embellished with
'off-the-peg' finials,
chamfered edges to
the rails and so on.

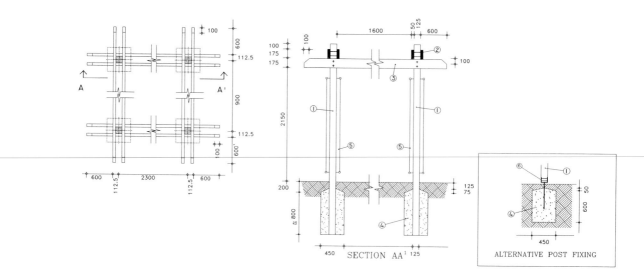

SECTION AA¹

ALTERNATIVE POST FIXING

Notes
1. 125 × 125 mm sawn softwood posts 3450 mm long.
2. 50 × 150 mm softwood spacer twice bolted to posts with 13 mm dia. galvanized coach bolts; countersunk.
3. 175 × 50 mm sawn softwood beams twice bolted to vertical posts with 13 mm dia. galvanized coach bolts.
4. Concrete foundations using C:20:P mix concrete.
5. Plastic-coated wire tied tightly to non-rust screws, top and bottom of post, 4 No. wires (8 screws) per post.
6. Galvanized metal support shoe of approved manufacture fixed to post 1 twice bolted with 2 No. 13 mm dia.coach bolts.

General notes
• All dimensions in millimetres
• Do not scale from this drawing
• All timber to be CCA treated to BS 4072 Part 1: 1987 'Tanalized' or similar and approved. Plus 2 coats of Sadolin 'Classic' walnut in accordance with manufacturer's instructions.
• Climbing shrubs to be positioned in accordance with scheduling near to wires and trained to wires with plastic-coated tags.

FIGURE 3.78 PERGOLA CONSTRUCTION DETAILS

duty garden wire must be tied tightly between eye bolts top and bottom of each side of each post, with further wires across the top of the pergola.

Arbours are round-topped tunnels of architectural metalwork, which can be of simple or ornate workmanship. The tunnel is really a grid of hoops and longitudinal braces, with further training wires for the twining of climbing plants. These can be obtained at relatively low cost from system manufacturers (such as Agriframe), though the most beautiful examples are always individually crafted.

Ponds

There are essentially four ways of constructing a pond satisfactorily:

1. The use of a pre-formed fibreglass liner, let into the ground with the edges concealed by York stone slabs or rockery stones.
2. The construction of concrete lined pools, with stone or slabs set over the rim to conceal the edges. Concrete is liable to crack and adequate movement joints are required, complete with rubber tie bars to seal the joints. A

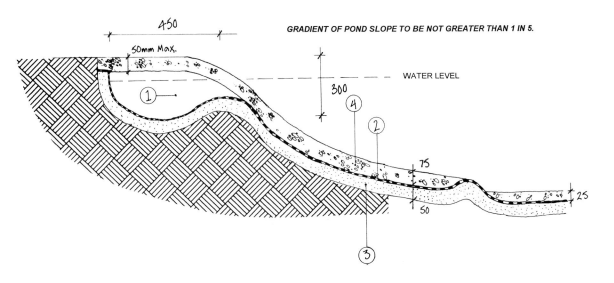

450

50mm Max.

GRADIENT OF POND SLOPE TO BE NOT GREATER THAN 1 IN 5.

WATER LEVEL

300

75

25

50

Notes
1. Topsoil infill as medium for bog plants.
2. Butyl pond liner.
3. Soft sand bed 50 mm depth.
4. 10 mm gauge washed rounded pea shingle, 75 mm depth
   as liner protection.

General notes
• All dimensions in millimetres
• Do not scale from this drawing

waterproofing compound must be added to the concrete and a sealant painted over the surface. Generally it is better to have vertical walls or the concrete will be too visible along the pond edges and the effect will be ruined.

3. The use of a pond liner to create a natural pond (see Figure 3.79). If the pond is sloped gently, with a ridge below the water line, then inert gravel can be laid over a protective layer of silver sand (avoiding the seepage caused by soil) and a natural beach-type edge can be achieved. Pockets of soil or pots of soil can be strategically located in order to establish emergent and aquatic plants. The liner should be butyl of sufficient weight to avoid puncture. It should be sandwiched in a layer of protective geotextile and bidim padding to avoid puncture from sharp stones on the pond bed. Herons can easily puncture liners and a layer of geotextile, then sand and then pea shingle over will minimize the risk of puncture in this way.

4. The use of a pond liner to create a formal pond (see Figure 3.80). Side walls are required of inexpensive dense masonry concrete blocks over a concrete foundation. To this perimeter wall is fixed an elm timber batten. The liner is fixed to this batten by screwing an elm lath to the batten, trapping the liner. The fixing is concealed from view by York stone slabs over-sailing the pond wall by 75 mm. If the batten is as near to the top of the wall as possible, then the final water level will be almost touching the base of the perimeter flagstones.

FIGURE 3.79 NATURAL POND CONSTRUCTION

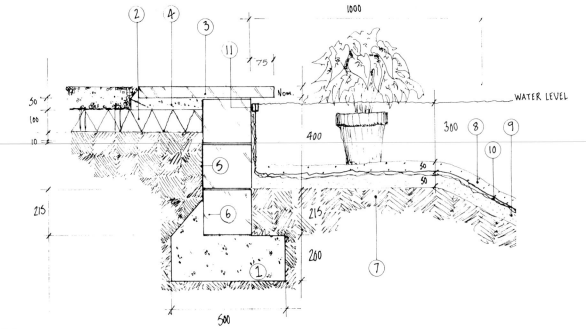

Notes

1. 500 × 200 mm strip foundation C:20:P mix concrete.
2. MOT Type 1 granular fill or similar and approved clean hardcore well consolidated to 75–100 mm depth to falls and cross falls as detailed.
3. Nominal size York stone slabs bedded, pointed and jointed on 1:3 mix cement mortar with 75 mm overhang to form edge of pond surround.
4. Cement mortar 1:3 mix.
5. 2 No. coats of Synthapruf bituminous paint.
6. 215 × 215 × 440 mm hollow block concrete wall blocks, jointed, pointed and bedded on 1:3 mix cement mortar.
7. Well-compacted subgrade.

8. Washed and rounded pea shingle 50 mm thickness to protect liner.
9. Sharp sand bed for pond liner.
10. Butyl pond liner.
11. PVC/elm wood batten and lath securely screwed to pond wall to secure pond liner or proprietary PVC alternative – to be approved.

General notes

- All dimensions in millimetres
- Do not scale from this drawing
- All timber to be CCA treated to BS 4072 Part 1: 1987 'Tanalized' or similar and approved

FIGURE 3.80 FORMAL POND CONSTRUCTION

It is essential to ensure that the following matters have been resolved before building a pond:

1. That there is no risk of the water flooding someone else's property, such as a basement, as the designer and client would be liable for the damage (strict liability as directed in *Rylands* v. *Fletcher*).
2. That the construction is entirely level.
3. That there is a method of filling the pond without the use of nitrate and chlorine rich tap water, such as a water butt fed from a roof gutter.
4. That there is an easy method of draining the pond, such as a plug leading to a drainage gulley.

5. That where there is a regular inflow there must be an overflow point with sufficient drainage capacity to absorb water precipitated by a 1 in 50 year storm.

6. That there is no risk of small children drowning in the pond.

CHAPTER 4

Soft landscape design

SOFT landscape is the term used for all living landscape elements such as grass, herbaceous and woody plants and trees. Soft works are those required to make living landscape elements grow and flourish in perpetuity.

In an introduction of this kind it is impossible to cover the full scope of this subject in all its immense detail; it has been well written about already and a wide range of literature is available on the subject. Instead this chapter suggests a rationale for coping with planting design which will serve to introduce the subject and help sort out the morass of information available.

Strategic landscape planting

To begin with, there is a distinction between strategic or masterplan planting and planting design of individual areas. Planners in local authorities will identify strategically important trees, groups of trees, woods and hedgerows, which contribute to the quality and character of the wider locality (see Figure 4.1). These can be protected by a specific mandatory protection order called a tree preservation order (TPO). A TPO is not required separately if the tree is lucky enough to grow in a designated Conservation Area. When a development is proposed, planners will impose landscape conditions on any outline planning permission granted. Such conditions will seek to preserve strategically important trees, tree groups and hedges regardless of whether or not they were already protected under TPO legislation. The planners may well seek such protection and their conditions will demand protection of the trees throughout the construction process and may also require enhancement of the trees or tree groups by additional tree planting. Such strategic planting will preserve the wider character and skyline of the neighbourhood and retain and enhance notable landscape landmarks in perpetuity.

The identification of such strategically important trees, tree groups, woods and hedgerows and the protection and enhancement of these features is one entire aspect of landscape planting design which embraces the disciplines of landscape assessment and landscape planning. This introduction, which cannot cover all the intricacies of the subject, serves to point out that these areas of the profession exist for further study.

STRATEGICALLY IMPORTANT TREE GROUP ON SKYLINE
AND SINGLE TREE IN MIDDLE DISTANCE.

FIGURE 4.1 STRATEGIC PLANTING. Strategically important trees, groups of trees, woods and hedgerows contribute to the quality and character of the wider locality.

Structural and ornamental planting

Structural and ornamental planting are terms that refer to the more intricate detail of specific locations within the site and are considered after strategically important existing landscape features and strategic planting have been considered. The landscape designer will concentrate on these individual areas, approaching them using the basic design principles used for landscape design in general, such as space creation and definition and so on. In applying such principles the planting will fall into either structural or ornamental categories. Structural planting can be summarized to mean space-creating and -shaping masses of foliage, which create a framework or structure to the site. Ornamental planting is the detail, adding individual character and distinctiveness by the use of colour, flower, texture, etc. These terms are explained in more detail in the following passages.

Enclosure of space: mass and form

Plant mass

One of the most important principles when approaching soft landscape design is space creation and plant form. Plants are materials that have a definable mass (see Figure 4.2). A group of plants of one species will present a mass or solid character of a given height and width. By bringing together a series of plant groups, you are able to arrange different masses to encircle spaces or unite different blocks to form the overall mass that defines and encircles the space itself. The bold use of planting to form space-defining masses is known as structural planting. The more intricate arrangement of plants against such structural masses can be defined as ornamental planting.

As outlined in previous chapters of this book, spaces are like outside rooms, which must be large enough and comfortable enough for the functions and activities (or inactivities) the client wishes to provide for. The size of each block of the enclosing planting mass will be determined by the numbers of plants in each group. The species of plant chosen will determine the eventual height and width of each mass enclosing the space.

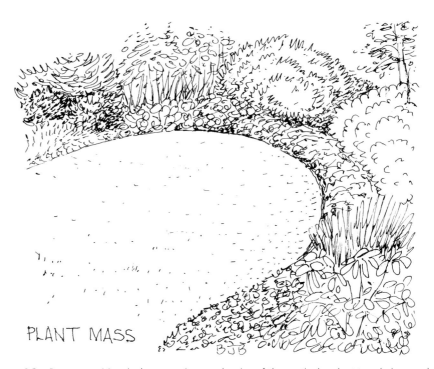

PLANT MASS

BJB

FIGURE 4.2   PLANT MASS. Mass is the term given to the size of the particular plant type being used, the bulk or space taken up by the plant or plant group. The further back the plant layer the larger its mass needs to be because the tall back layers are structural and the front layers decorative.

Plant form

Each variety of plant also has a definable form. When a group of one species is planted, the presence of this form becomes larger and more dominant. The form determines the character of a particular mass of foliage (see Figure 4.3).

Types of form

Some varieties will be 'procumbent', others 'erect' or 'fastigiate'. Some will be 'globular' while others will be 'pendulous' or of rectangular form. Many plants grow from a single stem and branch out in an up-ended shuttlecock effect. Others grow from the base and may sprawl along the ground or arch, while some shoot out spear-like leaves in a rosette pattern.

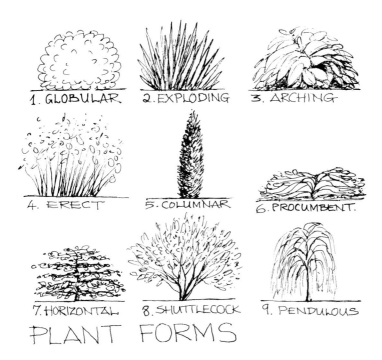

FIGURE 4.3   PLANT FORM. The many different plant forms should be juxtaposed and contrasted as much as possible to achieve the best effect. Form and habit are traditionally separated but when looking at individual plants it is usually one of the two that dominates the plant's character and determines how it should be used.

Using mass and form

Before you think in terms of individual varieties of plant, it is important to change your perception about plants. Think instead of plants as just another construction material, for the creation of space and subsequent modelling of the space and for ornamental relief. Think of this material as malleable, and varied, too, with a variety of mass and form. Think of the plants as blocks, pyramids, cylinders, shuttlecocks and globes. Then introduce some form to the shapes. Some will have a horizontal grain, others vertical, drooping, rounded or spreading. The form will be defined by the plant's habit, its shape and its texture. All of these together provide a complete character.

Habit is the direction and density of the branch structure. Texture is the term given to relative contrast between light and dark, that is, leaf size. Big leaves create the greatest contrast between the lightness of the leaves and the dark spaces between and therefore create the coarsest texture. Small leaves create lesser contrast and therefore present a finer texture.

With the concept in mind of plants being building materials, a plan or rendition can be drawn showing schematic plants only, purely chosen for their contribution to the tasks in hand. Actual plants can later be chosen to fit the intended characteristics. Often there are several varieties to fit a required form and manner.

Plant function

Plants can be positioned and grouped to carry out a variety of functions in the landscape. Plants can shelter (see Figure 4.4), screen, muffle, provide security, cleanse, cool, guide the eye, direct traffic and so on. Plants can be used to create an attractive composition (foreground, middle ground, background).

Functions can be divided into practical functions, such as screening or

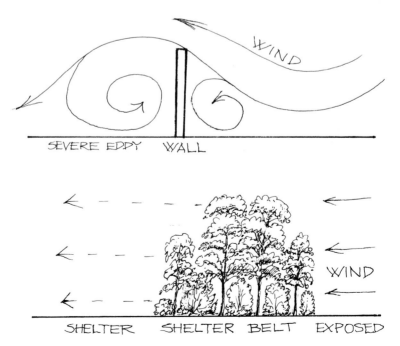

FIGURE 4.4   SHELTER. Whereas hard materials such as brick walls and timber fencing obstruct wind, causing eddying both in front and behind the obstruction with very little sheltering value, planting is semi-permeable and allows the wind to travel through, but it will slow the wind down dramatically creating a high degree of shelter. Any gaps, however, will cause funnelling and increase exposure.

provision of shelter, and aesthetic functions, such as defining spaces and providing focal points and ornamental groups. When using plants for their aesthetic function in landscape design there are three broad categories of use, as set out below (see Figure 4.5).

Structural

Structural planting belts and blocks define space, though crudely and simply. They tend to comprise tall, mainly evergreen (but also some deciduous) varieties of both plants and trees. Such plants are positioned to form the framework of the design. This framework will be a backcloth for the detailed ornamental planting arrangements.

A common backcloth material used in garden design is the traditional yew hedge. Although the yew is very slow growing and sometimes difficult to establish, the consistency and ease of maintenance that it provides make it the number one choice for providing a structural backcloth in the garden. It is, however, sometimes hard to persuade the client to be prepared to wait the length of time required to establish a dense yew hedge. Moreover, in this litigation-conscious time thought must be given to the liability factor. Yew seeds are extremely toxic and may pose a risk where small children might foreseeably play. Although a well-pruned hedge will minimize this risk, there may be no responsible grounds manager or gardener to oversee the work. There are, of course, alternative hedge materials such as Portuguese laurel, *Escallonia*, *Osmanthus*, some Cotoneasters, *Viburnum tinus*, Beech and Hornbeam and many others. Few hold their shape quite so well as yew although some of the shrubs mentioned above are better for low maintenance and informal hedgerows or backcloth shrubbery.

Ornamental groups

In front of the structural planting, other plants are used for ornamental groups and arrangements that provide the individual character of each space, and soften and embellish the structural planting. The arrangements are set against the backcloth of the structural planting but are of various heights, usually arranged approximately in bands from taller at the back to ground cover at the front. There will be some more prominent plants or plant groups as well as the 'fillers' of more bland plants, the latter acting as foils to the more showy varieties.

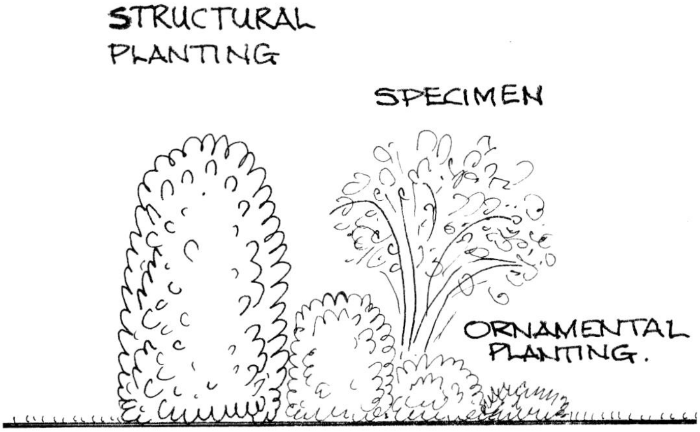

FIGURE 4.5 TYPES OF PLANTING
1. Structural planting will be tall, bold masses of mostly evergreen planting to create and define space.
2. Further layers of ornamental planting, which can be deciduous and include herbaceous material, are set against the structural planting.
3. Specimen plants are feature plants of architectural form or seasonal colour.

STRUCTURAL PLANTING

SPECIMEN

ORNAMENTAL PLANTING.

Specimen features

These features include focal points, accents and unusual or striking specimen plants (sometimes referred to as 'architectural plants'), which provide the punctuation in the landscape. They are used to improve the visual composition, for example to stop the eye along a vista perhaps, to catch the eye in an open space or to lead the eye from point to point. Such plants are usually of bold, architectural form and may be brightly coloured, very coarsely textured, of very large size or indeed whatever is necessary to contrast with the background ornamental arrangements and structural planting.

Planting philosophy

Context and mood

Plant context is a term meaning the flavour or character of the landscape governing plant choice or actually created by the particular choice of plants or indeed both. Such contexts might include formal, ornamental, natural, native or a particular habitat type, such as acid heath, woodland glade, waterside, meadow, downland and so on. There are many different cultural contexts to planting design too, such as Japanese, Moorish, English Romantic, Italian and so on. Context in planting design can refer to the mood created by the setting and how this setting is reflected in the choice of plants, whether grand or intimate, urban, suburban, rural, historic, modern, classical, romantic, cottagey, palatial and so on.

Plant layers

The conventional path to achieving a pleasing aesthetic effect when designing with plants is to mimic nature's pattern. In nature plants grow within the boundaries of their ecological niche; that is, plants will be associated with individual habitats and locations within these habitats, according to their preferred microclimatic and other environmental factors.

Some plants prefer a woodland setting, while others prefer the woodland edge or open meadow. Plants therefore can be identified in groups sharing a preference for particular locations and habitats. The height of a plant is a useful factor to identify. If you were to draw a cross-section of a wood and glade, you would find canopy trees, sub-canopy trees and climbers, shrub layer plants and ground layer plants and herbs.

When designing a landscape we often copy these naturally found layers in our planting to good effect, though the degree to which nature is abstracted is determined by the context of the site and brief. Mimicking nature in an abstract way produces formal planting, whereas mimicking nature as itself, using native species, gives a more natural feel (see Figure 4.6).

Abstract planting

Formality and order may be elements of the design proposals that demand very regulated use of plant layers. The crisp topiary columns may represent the canopy layer, while the sharp outlines of rectangular hedges represent the sub-canopy layers and the borders will house the shrub and ground layers, with the grass and herb layer represented with bright green, manicured lawns of finely mown grass. Ornamental species will be used to represent the native plants, which depict the various layers in the woodland, and may allow a variety of bizarre architectural forms, accents, features and contrasts not possible with the native plants (see Figures 4.7 and 4.8).

True to life planting

Local species will blend in naturally with the native woodland flora, especially if the varieties are copied in the proportions and groupings found in the particular area. A soft, natural undesigned appearance can result, which is usually the goal of such planting.

FIGURE 4.6 NATURAL PLANT LAYERS OF NATIVE AND SEMI-NATIVE PLANTS. Plants naturally grow where it best suits them ecologically: a wood or glade consists of canopy trees, sub-canopy trees and climbers, shrubs, and ground layer plants and herbs.

FIGURE 4.7
ORNAMENTAL ABSTRACT
PLANTING. Natural
height gradations are
imitated with non-
native plants ranging
from sub-canopy trees
to shrubs which grow
to different levels and
whose contrasting
forms provide strong
accents.

FIGURE 4.8 FORMAL
ABSTRACT PLANTING.
Formal abstract
planting can
incorporate topiaries
and hedges or walls
with arches to
represent the canopy
and sub-canopy layers,
with ground layer
plants represented by
bright green,
manicured lawns.

Natural habitats can be created from scratch by following the correct ground preparation to ensure that the soil is of a suitable type, nutrient status and pH value (pH is a measured scale of acidity or alkalinity). Introducing the correct mix of species for the various layers encountered naturally will ensure an appropriate plant association to encourage a diverse range of local fauna.

Various levels in between

The majority of planting falls into neither the entirely formal nor the entirely natural categories but is somewhere in between. Although there is some variance of commission according to geographic locality (National Parks and Areas of Outstanding Natural Beauty precipitating more natural planting schemes), the main source of commissioned planting schemes is centred on the need to provide for recreation and to mitigate the effects of industrial and residential development.

For general amenity planting of parks, public open spaces and gardens a higher proportion of ornamental species may be appropriate (but not exclusively so), as the aesthetic and activity functions are more important. The designs may be informal but are nevertheless of a more deliberate and controlled style.

Mimicking nature

By mimicking the natural layers formed by different plants that have adapted themselves for a specific niche in their habitat, a designer is able to make native planting schemes blend in with the natural vegetation. Even ornamental schemes will appear more attractive when plant layers are designed into the scheme. There are five naturally occurring layers:

The grass and ground herb layer

Visually, this layer provides the touchstone to measure the other layers by. It is the void; the space defined by the masses of the other layers. It is used by the client or end users for the functions intended and for wild flower meadows and glades.

Ground cover (shrub and herb) layer

This layer provides a carpet appearance for the fronts of beds, under large shading trees, and anchors leggy, 'shuttlecock' forming plant types and suppresses weeds. It is used for herbaceous borders including wild herbaceous woodland fringe or glade species.

FIGURE 4.9   PLANT ASSOCIATION. It is essential to contrast the primary design criteria of texture, character and colour of plants, but the best positions for their relative height (the other primary design criterion) and the way they grow must first be considered.

Shrub layer

This layer supplies the mass or lower level structural planting. It provides the feature and filler plants against the backcloth species, for softening structural planting and for anchoring leggy plants. In native planting schemes such plants might include *Cytisus praecox*, *Viburnum opulus*, *Ulex europaeus*, *Cornus sanguinea* and so on.

Sub-canopy layer

This is the layer for tall structural definition of space, for features and accents. It provides scale in spaces, casts dappled shade where required and offers shelter and screening. In a woodland such species would grow under the larger forest trees and would include hawthorn and hazel.

Canopy layer

This layer defines space in terms of height, casts shade and provides shelter in shelterbelts when combined with other plant layers. It supplies large-scale impact and a sense of scale in the landscape as a whole and contributes to the landscape character and skyline profile.

Factors affecting plant selection

There are so many plants to choose from and so many factors that must influence a designer's choice that it can appear a baffling and daunting task for the student and indeed even for the qualified designer. In order to sort out the morass of information and actually make a confident choice of plant, it is necessary to have a rationale, a guide that breaks up the information into chunks that can be more readily understood and used.

Narrowing down the choice

It is important to understand that any one variety of plant is just a tool or material, which simply performs a job, and its role is determined by its specific characteristics of form, mass, texture, colour, height and habit in relation and contrast to the plant nearest to it. This means that you may opt for one of many possible choices of plant, which might be equally suitable for the purpose intended. You may therefore be able to find a suitable substitute if a variety is not available in the required size or if you wish to keep to a restricted budget by using the lower priced options. For example, *Hypericum patulum* 'Hidcote' is a far cheaper shrub than, say, *Sorbaria sorbifolia* or *Choisya ternata*. Also, some plants might possess the right characteristics but may not be suitable for the specific site conditions, such as soil and microclimate (see Figure 4.9).

Plant menus

In order to decide on the actual plants, it is essential to develop menus of plants which are categorized into the groups summarized earlier:

1. Structural plants.
2. Ornamental (prominent or filler) plants.
3. Specimen features or accent plants.

It is a useful exercise to analyse the effectiveness of your designs or other people's designs in terms of the above categories, by looking at the choice of plants and thinking about what job they do and how they perform in relation to these categories.

You can expand these groups by providing sub-headings for the various aspects of each category, including mass, form, function, colour, texture, height and habit. Care must be taken with the use of these divisions as some plants may have an erect habit (in terms of their branching structure), yet their most striking characteristics are the rounded form and dense coarse-textured leaves. The habit will therefore be masked by these other features and as such is irrelevant for the purposes of plant choice. Indeed, it could actually hinder correct choice. It is therefore necessary to list only the prominent characteristics of the plant when preparing plant menus.

Every site you design will be slightly different from the one before but nevertheless there are many similarities that will make plant menus an essential time-saving tool. Organize master lists for various soil types, microclimatic harshness and so on and then rely on these basic lists to provide the essential functional principles successfully, again and again. New plants can be added to this master list and others reviewed and removed if necessary. Add site specific plants too and thus your knowledge and awareness of plants will increase. Above all you can make your planting designs more varied, subtle and interesting without compromising the fundamental principles of planting design.

Ground and site factors

The following ground and site factors should be considered when preparing a master list of suitable plants, and will allow many unsuitable plants to be rejected for each particular site. This rejection might be on the grounds of the plant's requirement for acid soil, its non-hardiness, requirement for full sun, over-vigorousness or drought intolerance.

*   Microclimate – hardiness, exposure tolerance
*   Soil type/pH – degree of tolerance
*   Shade/sun – tolerance
*   Damage/wear – durability/bounce back/wear tolerance
*   Climate – hardiness, altitude-hardiness
*   Vigour of growth – effect on surrounding plants
*   Moisture – requirements, tolerance

Outside factors

The following external factors are not related to basic horticultural concerns but are nevertheless common problems faced by designers.

Cost

The cost of each plant species is to some extent dependent upon the relative ease of propagation. Expensive shrubs include *Photinia*, *Mahonia*, *Elaeagnus*, *Garrya* and so on. Shrubs that are easy to propagate cost much less and such relatively low priced plants include *Rosmarinus*, *Cotoneaster*, *Hypericum*, *Vinca*, etc. It may be possible to substitute a cheaper variety of shrub and still retain the intended effect. *Cotoneaster lacteus* is similar in its functional characteristics to *Elaeagnus ebbingei* (being a medium-textured, tall-growing greyish green evergreen shrub of vigorous growth and wide site tolerance) but is half the price.

Availability

Availability is another factor that affects price, but more importantly plants that are hard to source may delay a contract, be substituted or be obtainable in an inappropriate size. Theft of plants is a common phenomenon, especially the attractive ornamental varieties, and may be a factor in the choice of bland shrubs in planting schemes in some areas.

Theft and vandalism

Theft is far less predictable than vandalism. Vandal prone areas can be anticipated to some extent. The use of tough and bland shrubs does help to mitigate the problem but at a cost to the attractiveness of the scheme.

Poison – dog and cat urine

Many plants are very sensitive to dog and cat urine and while these plants might grow perfectly well under any other circumstances, if they have been planted on the beat of a scent-marking pet they are likely to die suddenly, apparently for no good reason. Commonly specified plants that are affected include hebes, with 'Autumn Glory' being particularly prone.

Design criteria

This section is split into two parts, the primary and the secondary design factors. The primary design factors set out below are those that will need to be kept in the forefront of the designer's mind no matter how much knowledge about plants is acquired. Secondary design factors can be defined as knowledge of plants and their characteristics. With experience this knowledge becomes second nature so that the designer can concentrate on the arrangement of plants using the primary design factors. A useful analogy would be mowing a lawn. If the primary design factors equated to keeping a straight line and creating an attractive stripy lawn, then the secondary design factors would equate to the operating controls of the lawn mower. At first the operator will be struggling to learn the mechanics and controls of cutting, height, speed and so on, but once these are mastered he or she can concentrate on the effect.

Primary design factors

The primary design factors, as illustrated in Figure 4.10, are as follows:

- Height – low/medium/high: position in border (tall at the back graded down to short at the front)
- Colour – gold/green/purple/silver (the leaves only, avoid silver next to gold)
- Texture – leaf size: coarse, medium, fine
- Character – leaf/stem direction: erect, horizontal blob or hanging down (either the leaf or the stem will dominate to give character).

It is best to first think of the plants within their respective height bands. By doing this the designer has only to juggle three factors when juxtaposing plant types: texture, colour and character. The watchword with planting design is *contrast* (see Figure 4.11).

It is essential to contrast texture, character and colour of the plant, all simultaneously, to ensure that the plants are planted in the right position for their respective height. If you plant *Cornus alba* 'Spaethii', which has vertical stems, gold leaves and is of medium texture, in the medium height band, then a good plant in front of it in the low band would be *Cotoneaster horizontalis*,

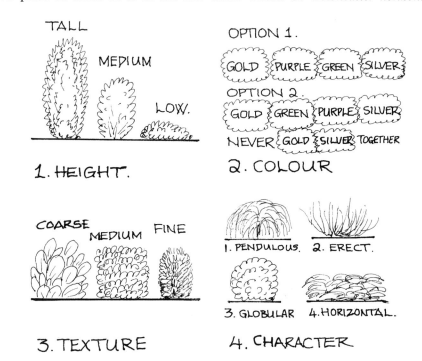

FIGURE 4.10 PRIMARY DESIGN CRITERIA. The most effective planting designs are those with plants selected to have the greatest contrast to the ones next to them. Contrast in height, colour, texture and character.

because it has a horizontal habit (when not against a wall), fine texture and green leaves. A plant to grow alongside the *Cornus alba* 'Spaethii' might be *Choisya ternata*, with its green leaves of finer texture and a flat or even-hanging character, which contrast well with the upright and golden-coloured *Cornus*.

Secondary design factors

The secondary factors listed below will be learned with experience and are important in selecting your plant list for the site, before you go on to position the plants.

- Growth rate/eventual size
- Flowers/fruit/seeds – seasonal variation
- Deciduous/evergreen
- Native/ornamental
- Ground covering or leggy
- Accent/filler

FIGURE 4.11
CONTRAST. Good contrast of texture, tone and character within any height band will stand out in black and white almost as well as it does in colour.

Eventually they will become second nature to the extent that you will not need to think about them.

It is important to ensure that there is a mix of evergreen and deciduous plants, to provide year-round interest as well as seasonal highlight. Also avoid putting slow-growing plants next to vigorous ones, or the vigorous plants will swamp the others before the slower plants have time to establish. Landscape designers should ensure that they choose sufficient plants with a good show of flower at various times of the year because this provides vital accents of colour and scent which people like. It is important that you choose plants that are of an appropriate character. Outside someone's front door, select ordered and tidy plants such as *Bergenia*, *Hebe* and *Potentilla*. Away from paths and dwellings or in country areas, looser, more natural looking plants such as *Symphoricarpus*, *Stephenandra*, *Ulex*, *Cornus*, *Sambucus*, *Rubus* and so on may be appropriate.

Some plants are natural accents, forming a shuttlecock shape, while others grow from the base and are useful to anchor the accent plants, hiding the skeletal stems that develop near to the base of the specimen.

Developing a plant directory

There are so many plants available to choose from that it is easy to become swamped and end up making bad choices. It is sound advice to develop a list of known and reliable plants and then use proven plant associations to contrast height, colour, habit and texture. Such a list can be expanded upon with experience and some experiment. Variation can be made to take into account the microclimate and position. In sheltered, well-tendered and small-scale gardens you can use more delicate and varied plant combinations. For large, exposed and infrequently maintained situations you will need more robust and vigorous plants to survive and create a satisfactory effect.

For example, some successful combinations might be *Prunus laurocerasus* 'Otto Luyken' and *Griselinia littoralis* behind *Vinca major* 'Variegata' and *Euphorbia robbiae*, respectively, which contrasts horizontal with upright habit, medium with fine texture and green with gold. *Rosmarinus officinalis* 'Miss Jessops Upright' behind *Artemesia arborescens* 'Powis Castle' is always attractive, but when *Rosmarinus* is set behind *Cotoneaster horizontalis* and with *Senecio dunedin* 'Sunshine' alongside, then there is a contrast of upright with horizontal and blob forms, fine with medium textures and green with glaucous and silver. In autumn the *Cotoneaster* leaves turn red and contrast well with the silver *Senecio*. In more exposed situations, such as those found adjacent to busier roads and where there is no shelter from cold winds, the *Choisya*, *Rosmarinus* and *Griselinia* could not be used, however. In the kinder, sheltered courtyards or with the protection of a south- or west-facing wall, perhaps the *Senecio* (for example) could be substituted for *Cistus salvifolius* 'Silver Pink' or *Convolvulus cneorum*.

Waterside and wetland planting

Plant layers

Just as there are natural plant layers in a woodland which designers can mimic in the design of ornamental planting so, too, there are naturally occurring layers by the waterside (see Figure 4.12). These layers are in fact particular ecological niches determined by the relative saturation of the ground. Some plants are adapted for a permanently aquatic life, while others like to grow with their roots under the water but their leaves above the water. Some plants prefer water-logged or marshy ground, while others prefer moist but not saturated soil adjacent to water. These plants divide into four groups:

- Aquatic – submerged: pondweed
- Aquatic – air breathing (rooted or free-floating)
- Emergent: plants with roots below the water and leaves above
- Marginal: marsh and bog plants.

Of course, there are some species that are happy to grow right across this spectrum, and such plants are useful for creating bold masses in large-scale water and wetland sites. However, for most small-scale and ornamental situations, it is preferable to use less invasive varieties that are more closely associated with one or two of these niches.

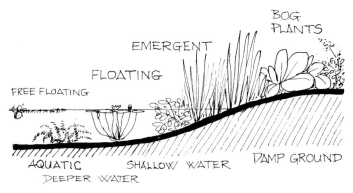

Value to the waterside habitat

There can be little doubt of the value of water associating plants, not only for their harbouring of aquatic and waterside insects, amphibians and mammals but also for their bank stabilizing value and their provision of shade and cover against predation. Many species are entirely dependent upon the natural plant layers for their life cycle: the dragon fly larva hides in wait for its prey in aquatic weed and then two or even three years later climbs up an emergent leaf to pupate and hatch into the insect imago.

Where erosion of the bank is occurring due to wave action, wash from boats or due to spate, then some species such as *Iris pseudacorus* are excellent at bank protection. Layers of silt washed down in spates can accumulate in the slacker water running through the leaves and actually build up banks. Indeed, dense growth of reed, rush and iris at the mouth of feeder streams to lakes and ponds can have several beneficial effects. The silt washing down the stream will eventually fill up the lake and it will become nothing more than a marsh. By trapping this silt in the dense growth, the sediment is deposited sooner and the lake will fill much more slowly. Where such feeder streams are polluted, large

FIGURE 4.12

WATERSIDE PLANTING. Plants can be chosen for particular niches in the aquatic and waterside habitats. Attractive waterside and pond planting utilizes plants for all the different layers shown above.

beds of rush and reed can act as giant water filter beds, trapping the harmful pollutants and cleansing the water entering the lake.

Such pollutants include the nitrates and phosphates that are the breakdown products of detergents and sewage, which are added to those already leaching into our water courses from agricultural land. When these nitrates and phosphates enter the water in quantity, the water can go green with algae. When these algae die, oxygen is removed from the water, killing fish and other aquatic life. The dead algae then coat the bottom killing oxygenating weed too. This process is called eutrophication. Banks of reed and iris that intercept pollutants and silt are sometimes referred to as 'buffer planting'. Since algae bloom in response to sunlight, some reduction of algae growth can be achieved by planting shade providing plants such as lilies. At the end of the day, it is far better to ensure that the water entering a pond is as free as possible from nitrate and phosphate contamination. This means using rain-water for ornamental ponds, perhaps diverting the guttering system of the building, and certainly tap water should never be used, even for topping up water levels, as it can be high in nitrates and phosphates and, moreover, contains chlorine.

Growing conditions

The main factor influencing which plants should be where is depth. Of course, some aggressive species can grow from depths of 1 m right up to dry land, for example the common reed, *Phragmites australis*. However, usually the plants have spread into such areas in times of low water level and many could not actually establish at depth. Indeed, many species, even aquatic submerged species, will not easily establish in deep water, which is too cold and sometimes too dark for them. Once established in warm shallow water, they will quickly spread to deeper water.

Some submerged species seem to thrive where there is annual draw down – the plants grow, flower and seed before the water level drops and then the seeds lie dormant on dry land until the following spring when water levels rise over them again. Marsh species, such as the marsh marigold (*Caltha palustris*), prefer waterlogged conditions and will tolerate a few inches of water for short periods. Some plants are oblivious to depth because they float, unrooted, for example frogbit (*Hydrocharis morsus-ranae*) and rigid hornwort (*Ceratophyllum demersum*).

Water flow is also important to plant establishment and survival. Spate rivers and streams, especially those with frequently shifting bottoms, may be able to support only mosses, while slower rivers will support a far wider range of plants. There is, however, no such thing as a nil flow water, though small garden ponds come about as close as you can get to it. The action of the sun and the wind will set up currents of surprising speed on lakes and ponds. Often the surface layers will be pushed along in one direction and an undertow set up below in the opposite direction. This action can also make the temperature stratification very uneven, especially in larger ponds and lakes, where warm water is pushed down into the depths on the lee shore. Here (usually on the north bank) submerged weed will be able to grow to a deeper level than on the other (usually the south) bank.

Another factor affecting plant choice will be the nature of the water to be planted, in terms of its cleanliness, clarity and pH. Plants such as the rigid hornwort are associated with alkaline water while plants such as the bulbous rush (*Juncus bulbosus*) are associated with acid water.

Few submerged species can cope with water containing much sediment for long periods because of the reduction in light. Pollution tolerance obviously depends upon the nature of the pollution – in terms of its toxicity to plants. Aquatic plants with floating leaves cope better than fully submerged species. Marginal and emergent species are less affected still and, as mentioned above, can have beneficial effects. Sewage fungus can coat the bottoms of polluted waters and this will smother and kill all aquatic life.

Design factors

The design criteria applied to terrestrial planting also apply to planting by or in water. Contrast is of prime importance, both in plant growth direction, colour of leaf, leaf texture and height. The only additional factor to consider is the proximity to the water. The absolute horizontal plane of the water surface can provide the perfect contrast to vertical waterside plants like *Iris laevigata*. Such plants can be planted beside or behind the more globular forms of the umbrella plant *Pelatiphyllum peltatum* or *Petasites*.

Again, as with terrestrial planting, there is a distinction between the controlled forms of ornamental planting and the wilder effect of native planting, and these should be chosen according to whichever style is appropriate to the context. There is nothing to prevent the gradual transition from the formal and ornamental to the informal and wholly native planting style.

Planting methods

Planting of aquatic and waterside plants is normally carried out in May, the first month that the plants become available from the nurseries, as this allows the entire summer for the plants to establish and grow. At a pinch they can, of course, be planted all through the summer, though late summer planting may allow insufficient time for the plant to establish, when such plants may be more easily lost in colder winters.

For all but free-floating plants, the rapid formation of a good root hold is essential to establishment, especially in faster-flowing water. For shallow emergent and marginal plants, notch planting or pit planting and firming in is the best method, but for all other plants growing in deeper water the plants must be attached to stones, planted in mesh containers filled with saturated soil and gravel or wrapped in a hessian sack containing stones. Like most plant material, the quicker it can be planted following its removal from the nursery, the better the plant will establish and grow. Water-loving plants are obviously more prone than most to drying out and mishandling.

Plant selection

Plant selection will depend upon the context of the site as to whether a native species or ornamental species is required, or even a blend of species. However, this book cannot possibly provide planting lists because of limited space. There are many specialist publications on planting design which can be researched and a site specific list can be prepared.

CHAPTER 5

The design process

The parties involved

THE job of landscape designer demands liaison with many different people, each with an important role in the building process. It is therefore essential that for designers to be professional, they must be familiar with the titles and roles of all the other parties involved. The parties that are likely to be encountered by a designer on a day-to-day basis are set out and explained below.

Client

The conventional arrangement in the landscape industry (not design and build) mimics that of the building world in that there are three independent parties involved, comprising the client, the designer and the contractor. The client requires a designer to prepare a survey, a design and then production drawings and documentation. The client will therefore have to commission this work, and in so doing a contract is formed between the client and the designer. On commissioning the designer the client agrees to pay fees (quoted in advance) for the service provided. The client may employ other consultants and, unless it is self-build, will later employ a contractor (on the advice of the designer) and pay the contractor for the work carried out (again on the advice of the designer).

Designer

Having agreed fees and terms, the designer collates client and site survey information and prepares a design. Additional services for the client might include submission of planning applications, estimating cost, selecting suitable contractors, tender action and administering the contract between the contractor and the client. Because the designer is responsible for the client's land and money professional indemnity insurance is required in case of errors or omissions that might invite litigation for breach of contract or professional negligence.

Contractor

Landscape contractors are quite simply builders of landscapes and may be selected from an approved list of such contractors by designers and invited to submit tenders for implementing design schemes. The successful contractor will try to maximize profitability on the job. Some contractors might try to reduce the specification or claim extras to improve profit margins. Regular inspection by the designer makes this more difficult. The contractor is paid by the client, often monthly. The designer will invite three or more firms to submit a tender price, then recommends the best tender (usually the lowest) for acceptance by the client, who then forms a contract with the successful contractor. The contract is a standard formal document, the purpose of which is to ensure that the client pays for the work that the contractor carries out, under the watchful eye of the designer, and that the work is up to standard and carried out on time.

Site agent

The site agent is the contractor's site manager, employed to manage and oversee the construction process and to order and check supplies and labour. The site agent will meet with the contract administrator for regular progress meetings and valuations.

Contract administrator

The contract administrator is usually the designer but may be a quantity surveyor. Acting for the client, the contract administrator inspects the works, issues variations and instructions, values the works carried out and issues certificates of progress and payment – up to final completion of the works.

Clerk of works

The clerk of works (C.o.W.) is employed by the client to check the construction on a daily basis, to check the quality and quantity of material supplies and workmanship and to inspect that both are in accordance with the specification. The C.o.W. will then report the findings back to the contract administrator.

Subcontractors

A subcontractor is any contractor who enters into a contract with another contractor, who is itself working for an employer. The contractor who has a contract directly with the employer is known as the main contractor. The contractor who has a contract with the main contractor is called a subcontractor. This contract between one contractor and another will usually occur because the main contractor is too busy to do the work which it is contractually obliged to carry out or because part of the work that the main contractor has contracted to do is of a specialist type, such as demolition, grading, architectural metalwork and timber fencing.

Domestic subcontractor

A domestic subcontractor is a firm contracted with the main contractor to carry out a service required in order to complete the job to specification. The main contractor chooses such subcontractors. Where the main contractor has made a decision to subcontract part of the work (assuming this is allowed under the main contract), then this subcontractor is called a domestic subcontractor. There is no contractual relationship between such a subcontractor and the employer (or the designer). It is the main contractor's responsibility to ensure that the work by the subcontractor is satisfactory. A main contractor usually chooses such subcontractors because they are competitive in price and/or reliable and do good work. The main contractor is liable for any defects made by its domestic subcontractors.

Nominated subcontractor

The term 'nominated' simply means that the subcontractor is specifically chosen by the designer and the main contractor cannot use an alternative subcontractor. A nominated subcontractor therefore is a firm employed under contract with the main contractor to carry out specific work. The reason why a designer might nominate a particular subcontractor will be because only that particular subcontractor provides the specialist and unique service required. No other subcontractor would be able to complete the job to specification. As with any option there are advantages and disadvantages with the nomination process.

Domestic supplier

A domestic supplier is a firm contracted with the main contractor to supply goods required in order to complete the job to specification. The contractor chooses such a supplier.

Nominated supplier

As with a nominated subcontractor, a nominated supplier is chosen specifically by the designer and will be employed by the contractor (under contract) to supply goods of a unique nature which are specific to the firm involved and which are required in order to complete the job to specification.

Quantity surveyor

The quantity surveyor is a consultant employed by the client to assist the designer with the preparation of bills of quantities for tender purposes and may also assist with tendering, contract administration and financial monitoring and control. (Bills of quantities is the term given to a document that contains in written form all the production information required by a landscape contractor to price, order and construct the scheme – see Chapter 10.)

Planning officer

A planning officer of a local district or borough council examines and either approves or returns plans and specifications submitted by designers seeking planning permission on behalf of their clients. Formal planning approval may be granted following negotiation and any landscape conditions imposed under the existing outline planning consent can then be formally released by the planning authority and notice of the same will be issued in writing, with copies of the plans stamped 'approved'.

The role of the landscape designer

One of the most interesting aspects of landscape design is the variety of work encountered, from tiny gardens requiring a single drawing (which provides all the information of hard and soft materials and specification notes – often the work to be implemented by the clients themselves) to huge development sites which require perhaps 15 separate drawings and a thick schedule of quantities. Such schemes may require liaison with all the people listed above.

The designer takes responsibility for someone else's land and money to fulfil their functional and aesthetic criteria in preparing design solutions, after taking full account of the site conditions, and the legal, planning and statutory factors.

The designer's role is a strange one, for there is no contract between the designer and the contractor; the designer has a contract only with the client. The designer administers the contract between client and contractor, being fair to both parties and arbitrating in any disagreement, after careful consideration of the facts. The designer has a contractual responsibility to the client to give best advice and because designers are liable for any negligent acts or omissions it is advisable to indemnify (that is, take out professional indemnity insurance) for the duty of care owed to the client in common law. The designer will need to act as the client's agent in all liaison with the local planning authority, statutory bodies, other professionals and in public consultation exercises.

It is important to mention at this point that the client will be called the employer in all documentation from the moment that a contract is formed between the client and the contractor. The designer has authority to administer the contract between client and contractor and to inspect the works. This is because the designer is acting on behalf of the client, as the client's expert – as his or her agent. Clearly, this role could provide cause for misunderstanding, confusion and dispute. For example, if the contractor were to seek advice directly from the employer (the client) and subsequently received conflicting advice from the designer (the client's agent), a dispute and/or additional cost could arise. It is easy to see how the contractor could act on an instruction given by the employer and then find that the work carried out affects foundations or causes some other structural problem and the designer has to ask that the work be undone. The contractor will be looking for payment for the work because it received an instruction from the employer. This may well increase the contract sum over budget if there are no contingency sums or if they have already been used.

Clearly, the designer cannot administer the contract and keep any kind of financial control unless he or she has the full and singular authority to instruct the contractor. It is imperative therefore that any instructions or advice that the employer wishes to give to the contractor must be imparted via the designer. The designer should be the sole voice of the employer and the contract between employer and contractor should spell this out clearly.

Where the role of the employer's agent (the designer) is set out clearly in the contract between employer and contractor, then should the contractor take instruction directly from the employer, it will be in defiance of the contract and such advice is taken at the contractor's own risk. No claim for additional expense incurred should be entertained if later the instruction cannot be endorsed by the designer. The contractor should never be encouraged to take instruction directly from the employer for, quite apart from the problem of technical fault or financial control, the authority of the designer is undermined by such dealings. Unfortunately, this phenomenon is not uncommon, not only in domestic garden design and residential development, but elsewhere too.

The advantages to the employer of the more traditional arrangement of using an independent designer are many. The employer can be sure that completely independent advice is received, with the most cost effective materials specified and only the most suitable plants chosen, rather than having to make do with those the contractor has left over and wishes to clear, or the materials for which the contractor receives the best margin of profit.

Design and build

A designer has two options: to be employed by or associated with a contractor jointly offering a design and build service or to be employed entirely independently, working directly for a client. Designers may find themselves working for many different people and it is increasingly common for freelance designers to work for landscape contractors on a design and build basis.

Design and build is the term used when the client approaches a contractor instead of a designer and then asks the contractor to design a scheme before implementing it. The fact that the contractor employs the designer means that the designer may not have direct contact with the client, acting instead on instructions given by the contractor. Such instructions might include insistence on the use of plants that are low cost or are currently growing in the contractor's nursery. Such curtailment of the designer's normal whims is sometimes frustrating but necessary if the design and build team is to make any money on a tightly costed commercial project. A designer may be given a fixed budget and have to calculate, for example, how many plants can be provided for the money. This kind of working relationship therefore brings an entirely different perspective to design.

CHAPTER 6

Client liaison, questionnaires and site appraisal

The exchange of letters

WHEN a landscape designer receives a letter (or phone call) from a potential client enquiring after design services, the designer will have to write (or phone) back to discuss the client's requirements and the services offered – when some indication may be given about general fee rates in order to maintain the contact. The designer should inform the enquirer that he or she cannot commit to a fixed charge until the site has been seen and a brief discussed. If the client is sufficiently interested to invite the designer to visit the site for an initial, investigative consultation to discuss ideas, an accurate fee quotation (or estimate) can be provided after the visit has been completed. It is strongly advised that the designer charges for the time and expenses incurred for such visits because unless these charges are levied the designer could waste a great amount of time, ideas and petrol on people who like to pick a designer's brains for free, generally waste the designer's time and who never intend to make a commission. A standard initial consultation fee can be quoted to potential clients to cover this stage of the proceedings.

The serious client will ask for a fee quotation (or estimate) and send details of the intended scheme, perhaps with a site plan, before a site visit is requested. The designer is then, on the basis of the information provided by the potential client, able to set out the services in greater detail (specific to the site) and provide a fee quotation (or estimate), payment terms and a programme of work. On receipt of this quotation (and no doubt others from other landscape designers), the potential client will write to the designer to accept or reject the fee proposal. If the letter accepts the fees, then this exchange of letters forms the contract between the designer and the new client.

Types of fee

To practise as a professional designer it is important to understand the different ways of charging the client for one's own work.

Percentage-based fees

Such fees are calculated on the basis of a percentage of the intended contract sum, in practice often a percentage of the budget for the job. A common problem is that after the design work has been carried out the client might decide to spend just half the money. It is therefore essential to quote on the basis of the contract sum or the budget, whichever is the greater sum. The percentage basis has recently become a less frequent method of fee quotation because most practices now work out how long the job will take and how many members of staff will be involved and so quote on a time or lump sum basis. However, it is still possible to work fees evaluated in this way back to a percentage figure, if the client requests this method. In the boom years of the late 1980s the percentage basis provided a good margin of profit which these days is more difficult to make. Nevertheless the percentage basis is still used where there is an identifiable contract sum.

The percentage quoted was at one time dictated by the Landscape Institute's mandatory fee scale, as set out in their document *The Conditions of Engagement and Professional Charges*. Perhaps unfortunately (at least for the practitioner) the Office of Fair Trading viewed this as a restrictive practice and abolished it. There is now an advisory publication by the Landscape Institute called *Engaging a Landscape Consultant: Guidance for Clients on Fees* (August 1996). This document sets out a very sophisticated methodology for calculating percentage-based fees, which involves a graph with job value on the vertical axis and percentage on the other. The smaller the job the higher the percentage. This percentage is qualified by complexity ratings applicable to different types of work according to how profitable they are. For example, private gardens are considered to be likely to cause the greatest hassle, requiring much client liaison time and many amendments to the scheme, so therefore a complexity rating of 4 is applied. However, road schemes attract a rating of 1 because of the repetitious nature and scale of the work. The fee is calculated by consulting the relevant graph for the particular complexity rating appropriate to the work in question, multiplying the percentage by the budget for the project. Of course, where the landscape architect is commissioned to provide a limited service then the fee is proportionately decreased. That is to say, the full service is deemed to be work stages C–L. Like all businesses, landscape practices have small and large commissions, with some more profitable than others.

Time-based fees

Under this system the designer's time is charged on an hourly basis at an hourly rate. A principal's time is dearer than that of other staff, who can be allocated a hierarchical rate. Although this makes the fee tender complicated by using three rates for one job, the client is assured of more competitive service. This saving results from ensuring that the 'boss' is not being paid (at the boss's rate) for work (such as draughting work) that could be carried out by staff, at lower rates.

The use of such a hierarchical structure helps to ensure that medium-sized (and to a lesser extent large) practices can offer an overall fee which, despite their higher overheads, is competitive with small or even one-person practices. Fee rates do vary a little from client to client as dearer fees will be charged to those companies or individuals where there is a likelihood of late (or non-) payment, to offset the cost of the loss of cash flow and either the loss of bank interest or indeed the payment of additional bank interest on overdrafts. Where many amendments are likely to be required and a fixed fee is requested, then clearly some account of the additional time must be made.

The hourly rate should be sufficient to pay for the designer's bills and overheads such as pension, mortgage, staff salaries, monthly accounts for business supplies, vehicle purchase and running costs, VAT, tax and annual expenses such as accountants' fees, Landscape Institute membership and practice registration fees. Maintenance and occasional upgrade of equipment should also be covered, as should any time off, sickness, client liaison and marketing time. There should be sufficient surplus as profit to add value to the practice and allow for investment in the practice's possible expansion or diversification.

The ceiling figure

A ceiling figure is used in association with time-based fees to provide an upper limit so that the client is not faced with a potentially unlimited fee bill. This is often necessary for the client to set a fee budget, but also provides some assurance that the job will not be allowed to take longer than it should. Clients do not like time-based fees for this reason, their worry often being that a job might take one practitioner eight hours, yet take another just four hours. They might also perceive that for a practice short of work there might be a temptation to 'spin out' a job. Such action is less likely than is feared as all practices wish to secure repeat business and therefore need to be (and need to be seen to be) both efficient and competitive.

The lump sum

A lump sum fee is a sum of money that the designer calculates will be sufficient to cover the total estimated time on the job. This fee can be paid in stages for larger commissions but for smaller ones is most often paid at the end of the job. The lump sum is undoubtedly the preferred fee basis of most clients.

For the larger jobs, which might include all the stages of work likely to be offered by the designer, it might take more than a year to finish the job, especially where contract administration is involved. Therefore stage payments will be agreed to coincide with the completion of the various parts or stages of the service. Perhaps 30 per cent of the total fee might be payable at final design stage, a further 30 per cent paid after the production of tender documentation, a further 10 per cent following tender action and the final 30 per cent payable at the end of the works contract, to cover the contract administration services.

Fixed price

This method means that the fee has been quoted in a tender situation and that the designer's fee quotation is fixed and will not be altered unless the scope of the work materially alters, for example extra work is commissioned, requiring additional design time. The fixed price ensures that the designer submits a fee quotation on a fair and equal basis, in competitive fee tender situations. Competitive fee estimates, even where there is a qualitative element to the bids, are nevertheless open invitations for a few less scrupulous practices to submit low fee bids, only to raise them after securing the contract. They might argue that they had underestimated the time taken to complete the job or that their quote did not include expenses or that an additional survey or other service is required where they can make up the profit margin.

Expenses

Expenses can be either included in the above fee charging methods or added to them. Expenses are always difficult to quantify in advance so it is common for rates to be quoted for such expenses, noting that such expenses will be in addition to the fees quoted. There is nothing sinister about this, so long as the quotation clearly states that incidental expenses are additional. It is helpful to set out a table of the charge rates for such likely expenses as car mileage, plan printing, copying and so on. Expenses charged should always be itemized and should be a true reflection of the costs incurred. Indeed, such expenses are mostly charged at cost of the nearest accepted figure. Mileage is charged at the Automobile Association rate applicable at the time. Other expenses include photocopies with different prices for the different paper sizes, including plan copies up to A0 size, dyeline prints, telephone and facsimile usage, photography, subsistence if on site all day, and normally any special postage or courier charges. There are, of course, other incidental costs such as copies of maps, local town plans and so on.

Terms

Rapid payment is crucial for cash flow but will not be offered unless terms and conditions clearly state when payment is due and even then rapid payment is rare. It is essential to state clearly all payment terms on fee quotations (and estimates) and on invoices. Insist on payment within 30 days and state that interest will be due on late payments.

Stages of work

Once the designer and client have agreed on the fee, the more detailed work can begin. The project then advances through the following stages.

Work stage A: inception
Brief, client questionnaire, site appraisal, budget and programme.

Work stage B: feasibility
Feasibility studies concerning planning issues, cost and technical achievability of objectives.

Work stage C: outline scheme
Transfer site assessment data on to plan form, using diagrammatic graphics, then proceed to outline proposals – overlaying assessment diagrams as necessary. Gauge cost.

Work stage D: sketch scheme
Work up the outline proposals on to larger-scale plan. Provide enough detail to enable the client to interpret and discuss the scheme. Amend as necessary and estimate cost.

Work stage E: final design
A plan that, following liaison with the client and other professionals (where involved), provides detailed information, well drawn and presented, with a detailed estimate of cost. Written approval by the client is necessary before proceeding to the next stages.

Work stage F: working drawings
Plans and sections showing the hard and soft construction, drawn in a conventional, technical manner for the contractor, including notes, schedules and specification data.

Work stage G: bills of quantities
Bills or schedules of quantities are documents bound in sequence that precisely specify the quality and the quantity of workmanship and materials of construction. The document is suitable for pricing by contractors invited to tender.

Work stage H: tender action
Prepare a list of suitable firms to build the scheme. Send them both plans and bills of quantities to price and submit by a due date. Report the results to the client, who then appoints a contractor.

Work stage J: contract preparation
Prepare the chosen standard form of building contract (e.g. JCLI form) and

arrange signing by both client and contractor. Provide any further production information required.

Work stage K: contract administration
Administer the contract. Report on progress and cost to client and liaise with clerk of works or quantity surveyor, if these consultants are involved.

Work stage L: completion
Administer the contract terms that relate to completion of the works and provide guidance to the client about the site after completion.

Other services
1. Preparation of public inquiry work including professional witness statements, proofs of evidence and attendance.
2. Landscape master planning, policies and strategies. All other work relating to planning issues and landscape designations, AONBs, etc.
3. Environmental impact assessment for major developments, roads, etc. that will materially affect the character of the landscape.
4. Landscape appraisal and evaluation for the protection of areas of special quality, prioritizing of funding or phasing of improvements.

Stages C to F in the aforementioned stages of work involve the production of drawings. Figure 6.1 will be of invaluable benefit to readers in summarizing the purpose, content and scale of these drawings.

Formulating the brief

Before pen can be put to paper, it is essential for the landscape designer to prepare (or to receive from the client) a thorough brief. The brief is a written statement of what the client wants the designer to do and should be as detailed as possible in order to ensure that the result suits the client's requirements. A brief should contain information about the client's needs and budget, and also about site constraints.

Client questionnaire and site checklist

There are therefore two entirely separate parts to the information gathering process for a landscape design project. The first concerns the client and the client's requirements, tastes, values, aspirations and budget. All this information can be listed on a standard client questionnaire form, which will be filled in by the designer at the first consultation meeting when the brief has been determined. The second part of the information gathering process involves the designer's own assessment of the site – which will be both by visual observation

FIGURE 6.1   THE WHAT, WHEN AND WHY OF DRAWINGS

Name	Purpose	Stage	Scale	Description
Survey	To provide an accurate base plan of site. For designer	Directly after briefing	1:100 to 1:500	A topographically accurate plan to show the existing site and items for retention/removal
Zone plans Type 1	Illustrate site analysis info. For designer	Early, after briefing	1:500	Diagrammatic plan to allow visualization of site factors, i.e. soil, microclimate
Zone plans Type 2	Form basic site concepts and solve site problems. For designer/client	Early, after briefing	1:500	Diagrammatic plan to allow visualization of first design ideas and principles
Sketch design	To crystallize first ideas into a co-ordinated design solution. For client	Midway following analysis/ concept	1:200	A rough drawn plan of the basic design layout, showing the spatial layout, all enclosing elements and traffic routes
Final design	To finalize the design ideas and work out the detail. To sell the scheme. For client	At the end of main design process	1:100	A worked up graphic presentation plan to sell the dream to the client, showing the final agreed layout of the site
Hard works layout plan	To provide a detailed working layout to enable the construction of the scheme. For contractor	Start of the production of working drawings	1:100	A simple layout plan unembellished using BS conventional graphics to show all structures/surfaces with levels and references to other drawings
Setting out plan	To provide dimension and measurements to enable the contractor to set out the layout on site. For contractor	One of the most important working drawings	1:100	A simple layout plan showing just the outline layout, with dimensions, running measurements, radii and offsets clearly setting out the scheme

FIGURE 6.1 (concluded)

Name	Purpose	Stage	Scale	Description
Demolition and dilapidations plan	To show all items to be demolished, excavated or else retained and made good. For contractor	Another working drawing. Midway	1:100	A simple layout plan drawn to show the items and areas to be removed/carted away and those that are to be retained and made good/repaired
Services plan	To show all services crossing the site – to prevent damage to them or accidents. For contractor	Another working drawing. Midway	1:100	A simple layout plan drawn to show the services crossing the site, such as gas, water, electricity, cable TV and telephone lines
Planting plan	To show the quantity, type, spacing and arrangement of plants and to schedule them for ordering. For contractor	Another working drawing. One of the last	1:100	A simple layout plan showing the outline of the planting beds with BS planting symbols, subdivided into blocks, each with the plant name, quantity and spec.
Construction details	To show the component parts of various elements to be constructed on site	The last working drawings	1:10 1:20	Unembellished plans, sections/elevations of detailed areas of construction, i.e. paving, walls, steps and fences, etc. to show how the parts all fit together

and by reference to the local planning authority, statutory undertakers (telephone, gas, electric and water companies), English Heritage (or equivalent), the Countryside Commission and any other relevant body. The client, neighbours and local council officers may be helpful with matters of ownership, planning legislation, rights of access, local flora and fauna, pests, etc. All such information is normally collected by the designer on a standard site checklist.

A typical landscape designer's client questionnaire and site checklist are set on pages 159–161.

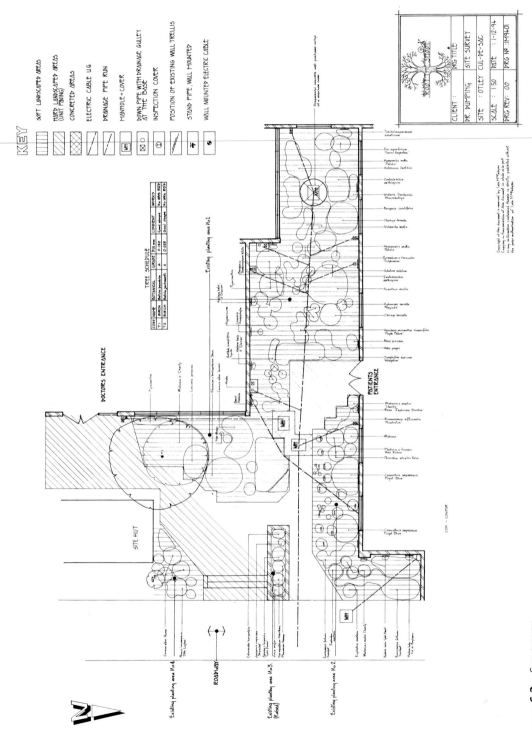

FIGURE 6.2   SITE SURVEY

Client questionnaire

Client group
1.   Age group.
2.   Health.
3.   Wealth.
4.   Children, ages, provision for.
5.   Pets, provision for.
6.   Tastes/values – likes/dislikes; style, plants, etc.

The client and the site
1.   Functional requirements: activities/inactivities – sports and ball games; sitting out; sunbathing; barbecuing; walking; entertaining, etc.
2.   Utilities: compost heaps, tool sheds, bin stores, incinerator area, drying area, vegetable garden.
3.   Mitigation of site problems: shade/shelter/privacy/security/screening/sound attenuation and so on.
4.   Enhancement of site assets required: framing views in and out of the site, clearing ponds, streams, etc.
5.   Desire lines/routes required/retained/re-routed.
6.   Artefacts/sculptures/focal points to be sited.

Maintenance
1.   Desire of client to avoid maintenance.
2.   Reliability of existing maintenance; gardener.
3.   Skill/knowledge of client/gardener.
4.   Budget for maintenance.

Budget
1.   Amount available/flexibility.
2.   Source: private/public/grant/charity, etc.
3.   Phasing – if any.
4.   Strings: contingency/fees/VAT included/excluded.

Fee arrangements (where not agreed)
1.   Percentage basis/time basis/lump sum, etc.
2.   Payment method/terms.

Site checklist

Survey – visual factors
1.   Climate: local, national, microclimate; degree of shelter/exposure; degree of sun/shade.
2.   Geology and soils: underlying rocks, soil type pH; shrinkability of soil and building foundation depths.
3.   Hydrology: drainage; water table.

4. Vegetation: habitat type, native plant types.
5. Existing trees (TPOs): name, size, grade, condition and works required to improve them.
6. Land use: proposed, existing, historical; adjacent.
7. Land ownership: boundaries – position/type; rights of access, service easements, wayleaves, etc.
8. Archaeological and local history: Biological Records Office; Sites of Special Archaeological Significance.
9. Period of building/surrounding buildings/gardens; style imposed: Japanese/ English/Italian/formal/natural/Romantic, etc.
10. Access/communications: pedestrian/vehicular; restrictions of access; rights of way; highways.
11. Pollution: water, air, noise, land, visual intrusion.

Planning matters
1. Town and Country Planning Act 1971: is a planning application required?
2. Proposed walls, fences, hedges and other boundary enclosures and their intended heights.
3. New and existing accesses, both vehicular and pedestrian.
4. Sight lines.
5. Visibility splays, turning areas and parking; structure plans/local plans, development and other zones.
6. Conservation areas, listed buildings.
7. Areas of Outstanding Natural Beauty (AONBs), Areas of Special Landscape Significance (AOSLSs), Sites of Special Scientific Interest (SSSIs), Sites of Special Archaeological Significance (SSASs) and so on.
8. Areas adoptable by county and district councils and commuted payments therein for ongoing maintenance.

Views into and out of the site
1. Vistas, views, focal points, avenues – framing views.
2. Composition.
3. Strategically important buildings, tree groups, hedges and woods.
4. Skyline profile/wider landscape character.

Site/land survey – levels and dimensions
Sometimes a site or land survey (see Figure 6.2) exists already, but where one does not it is usually more cost effective to use a specialist subconsultant. It takes half the time for a good surveyor to do the job better and the client would usually prefer the lower bill and the better job.

Mood and context
1. Urban/rural.
2. Formal/informal.
3. Grand/intimate scale.

4. Modern/traditional.
5. Contemplative/active.

The questionnaire/checklist must be used to ensure that you do not miss vital pieces of information, but do not get bogged down in trying to fill in the items for their own sake. Different items will be of primary importance for each site, while secondary items might need little recorded data.

It is important to prioritize items. Climate, microclimate and soils might have less relevance to a site where there is no soft works. Identify the features that are of primary importance for your site and circumstances. Highlight the primary items and these can then be lifted for incorporation into a site appraisal report.

The site appraisal report

Following the completion of the client questionnaire and the site checklist, sufficient information will have been gathered about the client's requirements (from the questionnaire) and about the site characteristics and constraints (from the checklist) to be able to compile a site appraisal report. A site appraisal report is a document sent to the client that presents all of the information hitherto gathered together and that sets out the main site assets, problems, functions required, the condition and value of existing features, and recommendations for these to be either retained and made good or else removed and replaced.

Industrial and professional bodies

There are a number of industrial and professional organizations that will be of invaluable use in obtaining design information. Whether it is the technical library of the Landscape Institute, the specialist advice of the Sports Turf Research Institute or the policies on public open spaces and play areas by the National Playing Fields Association – these bodies can be of great assistance in providing data at this early information gathering stage.

CHAPTER 7

Land survey

GATHERING information about the site and the client's wishes is, of course, a vital part of the design process but this information cannot generate drawings that accurately reflect the site unless there is a plan of the site, which depicts accurately the location of all existing site features. Land surveying (or topographical surveying as it is sometimes called) involves the accurate plotting of a piece of land to a recognized scale on a plan.

The main features of a land survey

(Figures 7.1 to 7.6 illustrate the points made in the text below.)

Boundaries and enclosing elements

It is essential to define on plan the legal land ownership boundaries and the means of containing them, whether hedge and ditch, fence, wall, bollards, line of trees or whatever, as well as any rights of access (for example for maintenance purposes).

Surfaces and changes of level

It is essential for the survey to show all paving types and soft surfacings, steps, kerbs, ramps, step ramps, retaining walls, revetments, piling, balconies, decking and so on. (A revetment is the term often used to describe the stone facing material on sloping embankments, such as for reservoirs and dams, to prevent erosion.)

Kerbs, edgings, channels, manholes and gratings

These items comprise, *inter alia*, pedestrian and vehicular division lines, edge restraints, surface water drainage patterns and service runs, with gulley location, type and size.

Levels and invert levels

It is vital when designing with hard materials, especially near buildings, to determine the precise height above sea level of each part of the site in order to ascertain gradients, proposed drainage patterns, quantities required for excavation and fill, the requirement for stepping walls and fences, and for ease of access. Buildings merit particular attention because it is essential to ensure that future surfacings do not bridge the damp proof course and it is always important in designing surfaces around buildings to prevent water falling towards a building, in order to avoid any risk of damp. Invert levels are the depths of inspection chambers and drain/cable runs, to determine how deep excavations can be carried out by machines without damaging services.

Soft landscape features

These features include tree girths, diameters, canopy height and spread, number of stems and the species of tree. The location of trees and shrubs is vital to show the position of sun and shade, to determine if any management works are required and to identify any site constraints. The accurate surveying of existing trees is necessary for most development proposals in order for planning permission to be obtained. The trees require detailed arboricultural assessment for this purpose too.

Services and service runs

The generic term 'services' is used to include water mains, surface and foul water drainage, electricity cables, transformers, junction boxes, telephone wires, gas mains and cable TV. Associated with the passage of services will be gulleys, grilles, manholes, inspection chambers, rodding eyes, stopcocks and gratings, and all such items must be plotted and labelled, along with contours, line types and representation of detail.

The basic principles of land surveying (*Geoffrey Yates of Sunshine Surveys, Suffolk*)

It is not easy for those inspired by the artistic creation of horticultural dreams to be riveted by the intricate trigonometrical theories that are required by surveying but it is, of course, important to get things in the right place or horticultural dreams will turn into litigious nightmares.

The following passages explain what is meant by the term 'surveying' in the context of landscape design. Generally for landscape works there is no need to delve into detail about satellite ground positioning systems and more advanced geodetic surveying as the areas of land under consideration are usually small. The following passages are entirely concerned with obtaining an accurate plan at a suitable scale, presented on a stable material and containing the correct information, in the most expedient way possible.

The detailed measurement survey

In the context of surveying for the purposes of landscape design, there are two realistic ways of ascertaining the size and shape of a piece of land (your site), which is likely to vary in size from several acres to just a few square yards. These are a framework survey and a traverse survey, as discussed below.

The framework survey

The framework survey involves the use of a 30 m tape. The convention of using metres in surveying was established when the Ordnance Survey chose to convert the County Series maps to a National Grid at the end of the last century. The grid was based on a kilometre as a unit and a metre as the standard subdivision. To convert measurements:

1.   Metres to feet – multiply by 3.2810.
2.   Yards to metres – multiply by 0.9144.
3.   Acres to hectares – multiply by 0.4047.

The essential point of a framework survey is to form a triangular frame and then to hang the detail from it. When growing runner beans, it is necessary to place strong and usually straight canes into the ground (pointing in the right direction) so the plants will be controlled in the way you wish. Similarly, land surveyors construct a series of triangles to form the frame. They ideally choose triangles to measure that are near to the boundary of the land or past important main features, so that they can see between the intersections of the sides. These are the 'stations', which are usually a nail in the ground or a wooden stake with a nail in the top. The surveyor first selects these positions and then commences measurement. The sides are measured with a tape in the field so that a scaled version can be drawn on paper to represent the same triangles ('in the field' being surveyor-speak for 'being outdoors'). To get more technical, trilateration means the measurement of three sides of a triangle and this is how the framework can be constructed. Figure 7.1 illustrates a plan of a house and grounds and how the picture can be built. Looking at Figure 7.2, the lengths AB, BC and CA are measured during the course of the survey when establishing the 'framework'. AB is drawn to scale on film or paper and arcs are drawn to scale from this baseline representing BC and CA.

However many triangles are required, it is essential to work from the outside inwards as to start from a small baseline or triangle and work outwards simply multiplies any inaccuracies. It should be noted that the plot of land in Figure 7.1 may fall into two halves by being divided by a natural feature. The survey should cover all the land to be surveyed and the features added to an overall framework rather than trying to treat them separately and then join them together. Now that the canes are in place, it is time to plant the runner beans . . .

The existing trees, fences, tracks and buildings are added to the framework by noting their distance along the line when measuring from station to station and the offset distance from the feature to the line. The term 'offset' just

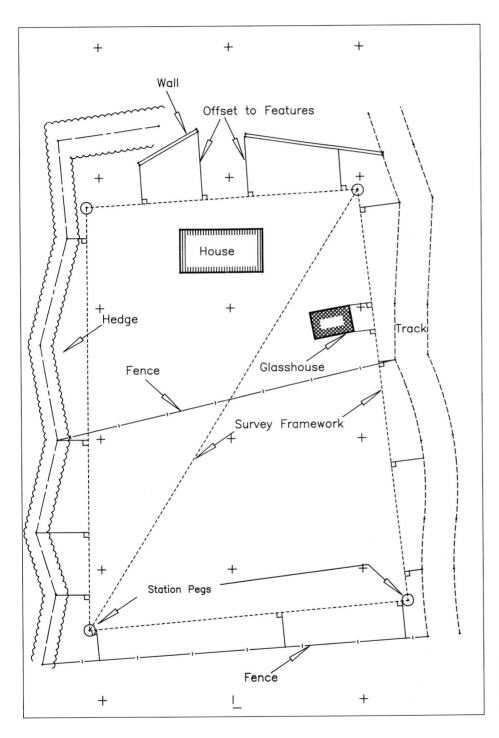

Wall

Offset to Features

House

Hedge

Fence

Glasshouse

Track

Survey Framework

Station Pegs

Fence

FIGURE 7.1
FRAMEWORK SURVEY —
PLAN OF HOUSE AND
GROUNDS

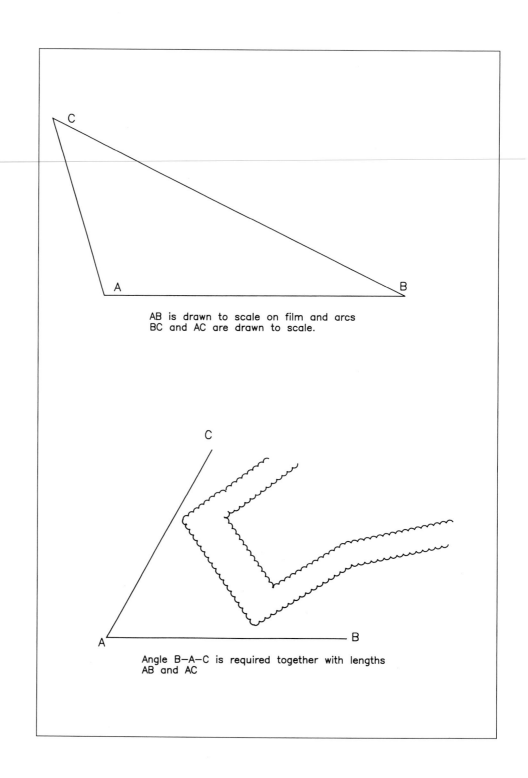

AB is drawn to scale on film and arcs
BC and AC are drawn to scale.

Angle B–A–C is required together with lengths
AB and AC

FIGURE 7.2
TRILATERATION TO
ESTABLISH THE
FRAMEWORK

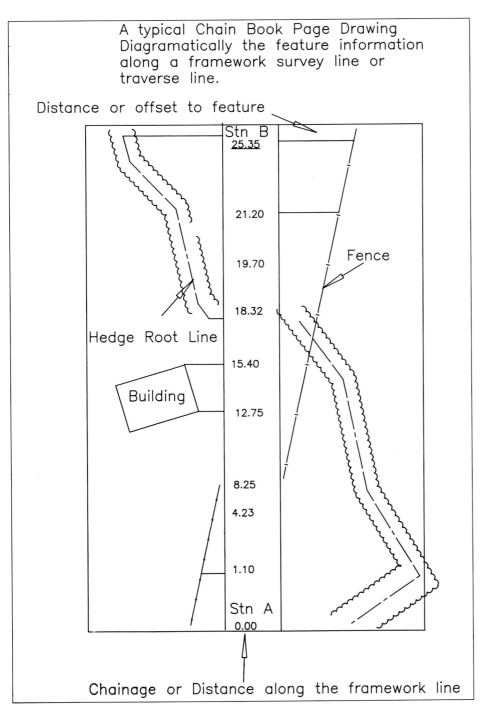

A typical Chain Book Page Drawing Diagramatically the feature information along a framework survey line or traverse line.

Distance or offset to feature

Stn B
25.35

21.20

19.70

18.32

Hedge Root Line

15.40

Building

12.75

Fence

8.25

4.23

1.10

Stn A
0.00

Chainage or Distance along the framework line

FIGURE 7.3
TYPICAL PAGE OF A
CHAIN BOOK

means the distance from the framework line, but it must be measured at right angles to the line and is then drawn at right angles on the survey plan. All this may be recorded in a 'chain book' and an example of the layout of a typical page is illustrated in Figure 7.3. The chain or tape in our case is represented by the parallel lines up the centre of the page. The surveyor will note the chainage (or distance) as he or she progresses and then write in the offsets, left and right to the features, as the survey proceeds. The stations are at the beginning and end of the line, the last measurement being the distance between the two stations.

The traverse survey

For more complex shapes, it may be necessary to start to measure angles and construct traverses. A traverse is the measurement of two lines and the angle contained between them (as noted in Figure 7.4 and subsequently in Figure 7.5 which increases the number of stations on a traverse). The use of geometry is a little more complicated: reference to the survey books is necessary if a full understanding is required. The principle is the same, however, and offsetting is carried out in the same way once the framework is established.

The typical feature details that are essential when building the survey plan are shown as a reminder in Figure 7.6. All these items must be surveyed and subsequently drawn on to the survey plan.

Delegation

The advent of electronic distance measuring instruments and computer-aided design (CAD) means that all the detail needed can be acquired more cost effectively by employing a land surveyor with this equipment. It is essential that the surveyor understands your requirements and has a good rapport with you. With modern equipment the surveyor will quickly be able to measure directly to all features by reading to a prism placed near to the feature to be surveyed, and transfer the measurements to a data collector, together with a full description and any appropriate remarks. The full survey data will be loaded on to a computer and, with a bit of editing, a drawing may be produced on a plotter, which takes data directly from the computer. Under these circumstances, the real effort must be applied to making sure that the land surveyor is given an accurate brief including details of the exact extent of the survey.

The involvement of a subconsultant surveyor using the CAD method has reversed the once certain fact that it was cheaper (if less accurate) to survey a site yourself. The few surveyors still using the old technology may well still be more expensive but it will almost certainly be less expensive to employ a surveyor using CAD than to go out yourself with a 30 m tape and surveying book and then draw up the plan later by hand – and far more accurate too. This will not, of course, be the case if the instruction is not clear and a repeat trip is subsequently required by the surveying professional, whether or not armed with the latest technology.

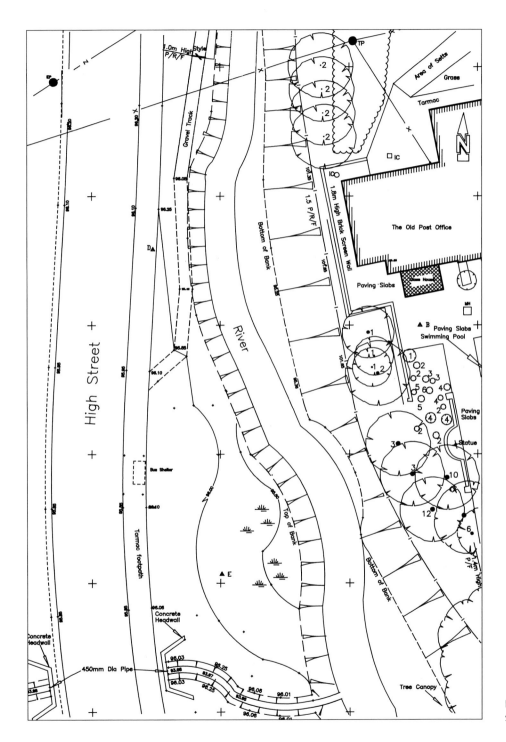

FIGURE 7.4 Traverse
SURVEY MEASUREMENT

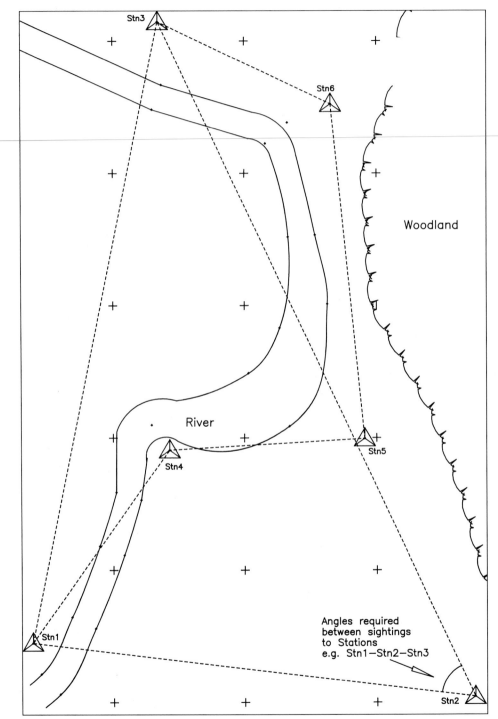

Woodland

River

Angles required
between sightings
to Stations
e.g. Stn1–Stn2–Stn3

Stn3

Stn6

Stn5

Stn4

Stn1

Stn2

FIGURE 7.5 TRAVERSE
SURVEY

Legend

Service Features:

AV	Air Valve	ER	Earth Rod	PM	Parking Meter
BB	Belishia Beacon	FDP	Feeder Pillar	RE	Rodding Eye
BD	Bollard	FH	Fire Hydrant	RP	Reflective Post
BM	Bench Mark	FS	Flag Staff	RS	Road Sign
BN	Bin	FW	Foul Water	SC	Stop Cock
BP	Boundary Peg	GP	Gate Post	SN	Sign Post
BS	Bus Stop	GV	Gas Valve	ST	Tree Stump
BT	Telecom Cover	GY	Gully	SV	Stop Valve
BTP	Telecom Pillar	IC	Inspection Cover	SW	Surface Water
CB	Cycle Barrier	IL	Invert Level	TAP	Water Tap
CC	Coal Chute	KO	Kerb Outlet	TB	Traffic Bollard
CL	Cover Level	LB	Letter Box	TF	Traffic Lights
CP	Catch Pit	LP	Lamp Post	TK	Tank
CT	Cable TV Cover	MH	Manhole	TP	Telephone Pole
CTP	Cable TV Pillar	MP	Marker Peg	TW	Telehone Wire
DPC	Damp Proof Course	MS	Mile Stone	U/G	Underground
EC	Electric Cover	NB	Notice Board	WC	Water Cover
ECB	Electric Control Box	NP	Name Plate	WM	Water Meter
EP	Electricity Pole	O/H	Over Head	WO	Wash Out
EPL	Electric Power Line	PE	Pipe	∅	Diameter

Hatching:

Building ;

Building/Shed

Glass House :

Building Surveys:

BL	Beam Level
BIG	Building Inlet Gully
CG	Ceiling Level
EL	Eaves Level
FL	Floor Level
HL	Head Level
RL	Roof Level
RWP	Rain Water Pipe
SL	Sill Level
VP	Vent Pipe

Fence Types:

B.W.F	Barbed Wire
C.B.F	Close Board
C.L.F	Chain Link
C.P.F	Chestnut Paling
C.I.F	Corrugated Iron
M.F	Mesh
P.F	Panel
P.W.F	Post & Wire
P.R.F	Post & Rail
P.K.F	Picket
W.I.F	Wrought Iron

Symbols:

Manhole/Inspection Chamber :

Inlet Pipe

Inlet Pipe — Out Flow Pipe

Control Station :

Bush/Shrub : Sp:3.0 Ht:5.0

Tree : Sp:8.0 Gr:0.6 Ht:12.0

Tree (Multi-Bole) : Sp:12.0 Gr:1.1 Ht:12.0

Gates : Single
Double
Style

Wet/Marsh Land :

Line Types:

Embankment : Top of Bank / Bottom of Bank

Telephone Line : U/G —×— O/H

Electricity Power Line : U/G —N— O/H

Drop Kerb :

Contours : —— 1 —— / —— 0.5 ——

Hedge :

Root Line :

Foliage

Tree Canopy :

Fence : —I———I—

FIGURE 7.6 SURVEY PLAN FEATURE DETAILS

After the surveyor's site visit, the survey drawing will then be produced on film, to the appropriate scale and, if included in the instruction, a number of plans can be produced from the computer data as set out below.

Survey drawing types and scales

- Location plan – 1:1250 – site in relation to local landmarks
- Overall plan – 1:500 – masterplan of whole site
- Detailed plan – 1:100 – large-scale plan of part of the site
- Specific plan – 1:50 – very large-scale detail of one area

Specific plans show the proportions of, say, a courtyard and any features within it, and such plans can be produced at up to 1:10 scale. The plans can invariably be arranged in the computer so that different features have their own 'layer'. This means that if a plot is required, the surveyor's text or existing buildings on site, or whatever is not required by the specifier, can be omitted. You may wish to have a base plan (this will just be a simplified layout of the whole site or part of the site), excluding all text but plotting existing buildings and adding your own logo and legend. The possibilities are vast but remember, if you use your time most efficiently, and use a surveyor with the right equipment, the client will probably be happier with a better surveying bill, and the designer will be happier with a better base plan.

Surveying is a separate and specialist area of work but, perhaps unlike many of the other professions that interact with landscape design, it is more fundamental to the process of landscape design. It is therefore often appropriate to employ the specialist services of a land surveyor at the outset of a landscape design commission. The surveyor may be employed directly by a client or subcontracted and paid by the landscape architect. Such fees can then be invoiced to the client as a disbursement with the landscape design fees. There is usually a 10 per cent mark up added to subcontract consultants' fees to compensate for the risk. The risk is that the client would take legal action against the designer (not the surveyor) in the event of errors or omissions in the survey work, as the client's contract is with the designer and not the surveyor. The safest option is therefore for the client to appoint and pay the surveyor directly.

CHAPTER 8

Evaluating and using the site information

Interpretation and implications

USING the site appraisal report and the land survey, designers will set out in note form or by using simple diagrams their interpretation of the data and resulting implications. For example, waterlogged ground may be either drained to make it dry or utilized for bog loving plants, or perhaps both in different locations. A harsh and exposed microclimate may require shelter planting, but this may obscure important views. A designer will highlight any such conflicts so that clients can make a choice according to their priorities. A client's desire for increased privacy and security on an open site may require the need for a planning application to erect a high fence. Permission may be refused and the fence may have to be modified, reducing its value in performing the principal function. Thinking through the implications of the data collection is a vital part of the design process.

Failure by the designer to see the implications of, say, the need to accommodate three cars in a garden and turn them comfortably will considerably waste the designer's time and a client will be unwilling to pay for such wasted time. If a client does not spot the oversight (being unused to reading plans) and in haste neither does the designer, the garden could be constructed only for the client to find that it does not provide for three cars parking and turning as required. The result may well be a writ served on the designer for professional negligence.

From data to design

There is a natural progression from data to design: from site appraisal to site interpretation (and analysis) – from interpretation to concept – from concept to sketch design – from sketch design to final design drawing.

Design is about synthesis – synthesizing aesthetic principles with materials of construction in a way that satisfies the functions and utilities needed by the client. Even from the concept stage, materials should be considered to some degree, taking account of their appearance, durability, cost and availability.

The designer will start with the basic elements, which put together will form a concept. These basic elements are the identification of suitable functions and moods, the choice of materials and the application of aesthetic principles. A good concept is founded on a good understanding of the client and site data. Good design arises from a sound concept. Design is about solving functional problems, and creating spaces for desired functions, utilities and activities to take place in. Identifying the context of the existing landscape and assessing the desired context is essential for good design.

The concept document

Following the interpretation and analysis of the survey data, a concept document is produced to set out the broad principles of the scheme. It provides the following details:

- The intended moods and the solutions to the site's functional problems
- The sizes, shapes and boundaries of spaces for the proposed functions and activities, along with broad statements about the desired character of the spaces
- The broad outline of structural and strategic planting, and aesthetic enhancement of existing features/assets
- The main intended materials of construction

Assessing function, mood and context

The designer needs a clear and rapid method to assess context and mood in order to ensure that the design works. Set out below is a simple methodology for moving through this minefield into more practical design considerations. Unless a clear picture is obtained about the nature of the context, the more practical matters of materials and spaces will be badly founded.

Identify 'key' words that sum up both the functions and moods required of the site, its desired character. For example, privacy, security, community, intimacy of scale, friendliness, comfort and relaxation are all positive words we associate with the context of 'home'. Lively, buzzing, fun, sociable, grand scale and colourful would all be words that ideally sum up the context of 'urban squares'. Determining the key words is the first step to establishing a brief for the site.

Once the designer has compiled a list of key words, then beside each key word (leaving a gap of about 50 mm) the word expressing the exact opposite should be written, for example:

Grand scale ——————————————————————— Intimate scale

Imagine a list of these key words and their opposites set out on a page with a line drawn (filling the gap) between them, evenly divided up as a scale from 1 to 10, for example:

Private	—1—2—3—4—5—6—7—8—9—10—	Public
Security	—1—2—3—4—5—6—7—8—9—10—	Adventure
Relaxation	—1—2—3—4—5—6—7—8—9—10—	Lively
Contemplative	—1—2—3—4—5—6—7—8—9—10—	Fun
High tech	—1—2—3—4—5—6—7—8—9—10—	Traditional
Formal	—1—2—3—4—5—6—7—8—9—10—	Natural
Classical	—1—2—3—4—5—6—7—8—9—10—	Romantic
Immediate	—1—2—3—4—5—6—7—8—9—10—	Mysterious
Serious	—1—2—3—4—5—6—7—8—9—10—	Whimsical

The designer can now assess the site, as it will be, by moving an imaginary pointer along the scale.

Once the context of the site has been defined using the key word scale diagrams, the designer can work out solutions to the functional problems and needs. The decisions about which shapes, features, materials and structures are appropriate will now be much easier. The choice will be that which is most expressive of the relative position on the scale for each key word. In other words, all the hard decisions about the scale, shape, style and character of the spaces are made in advance using this method. There is no need to agonize over the style of design, whether a line should be curvilinear or articulated, or a space large or small, as the designer can simply refer to the diagrams. This approach has a secondary benefit of starting the design process on a visual, graphic basis at an earlier stage, which further speeds decision making and makes it more certain.

Sketches

It is essential when first considering the design of a new site to make plenty of notes and to sketch some ideas at once in plan and perspective form in order to 'break the ice'. The longer this is delayed, the greater the chance of 'white paper shock syndrome', which can inhibit even the experienced designer from making any mark on the paper. Notes and sketches focus attention on problem solving. You may reject all of your initial ideas later as you compare your ideas with the brief, but at least the process has started.

Once started the ideas will flow and the next hurdle is ruthlessly to select those ideas that are most suited to the brief and co-ordinate them into a unified design. To achieve this unity choose a spatial framework to contain your ideas. A spatial framework that harmonizes and complements the features and ideas will ensure that they are as interdependent as a door is to its doorframe.

Zone plans

So what is the next step? The client's requirements and site assessment data can be shown graphically too by using small-scale (1:500–1:200) diagrammatic plans of the site – called zone plans – and such factors as the microclimatic variations, sun and shade and soil types/moisture content can be shown. The required activity functions can also be located on such diagrams and suitable-sized spaces allocated for them.

Zone plans are rough outline plans of the site, each illustrating a site factor of crucial importance to the design (see Figure 8.1). One set of plans may show existing factors such as the microclimates within the site, views in and out of the site and so on. Another set can be produced – bearing in mind the information about the existing site factors – for the proposed elements. If a zone plan is produced for the intended functions, according to suitable microclimate, aspect, distance from the buildings and so on, then a further plan can show the space sizes needed for the located functions and activities. Another zone plan might show the intended moods required throughout the site, while another will show the pedestrian and vehicular circulation (existing and proposed). Yet another is needed to show the structural planting between the spaces proposed, and this can be an overlay of the others. This zone plan can be produced and refined several times in order to begin to shape and reshape the spaces and to unify them as one whole design. Any initial ideas, concept drawings and overlaid zone plan diagrams are generally nicknamed 'fag packet' sketches. These 'fag packet' sketches, also sometimes referred to as 'back-of-an-envelope' plans, which show the first broad-brush ideas, can be drawn up on the precisely scaled base plan.

By overlaying all the zone plans and enlarging the plan to a suitable scale, the first bones of a sketch design will have emerged, which can then be refined further, providing the first indications of path widths, surface materials, focal points and features. The size, shaping, style, materials and features of the spaces can be developed to ensure that they perform the intended function and express the appropriate mood.

The designer will draught a base plan from the topographical survey and erase any existing site features that will not be retained. All the zone plans (showing the outline proposals) must be combined on to an accurate scaled up base plan of the site. This information can be added layer by layer, drafting spaces for the functions required, the structural and strategic vegetation used to define the spaces, the desire lines and routes for both pedestrian and vehicular circulation, the aesthetic factors and the materials of construction.

The process of crystallization

During the process of sketching ideas, making written notes and designing the spatial framework, a process of continuous crystallization is taking place, where

ZONE PLAN - MICROCLIMATE.

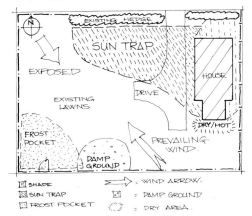

ZONE PLAN - SPACE AND MASS

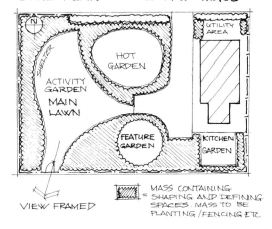

ZONE PLAN - CIRCULATION

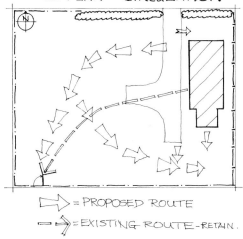

ZONE PLAN - OVERLAY

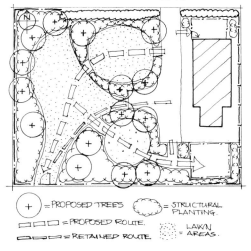

FIGURE 8.1 ZONE PLANS

1. Microclimate – this is a small base plan showing the existing survey features of the site on to which are drawn the salient points of factors that will affect the microclimate in different parts of the site, e.g. exposed winds, sun traps, frost pockets and so on. These factors will determine the location and extent of activities and the choice of hard and soft materials.

2. Space and mass – the intended activity or use will be positioned according to the microclimate assessment and a space allocated of sufficient size for the activity or use envisaged (size determined by aesthetic and practical factors). The space will be defined by enclosing it and creating

substantial masses around the spaces, which should be linked together to unify the spaces with site boundaries. Much of this mass infill material will probably be planting although other enclosing elements such as walls, fences, trellis work and so on might be included.

3. Circulation – existing pedestrian and vehicular routes will be graphically depicted along with proposed routes to provide a clear picture of where roads and paths have been located to take account of desire lines and established routes.

4. Overlay – the circulation can be overlaid with the spatial diagram and further defined with choice of materials for the surfaces and mass space-defining materials to form a concept plan.

thoughts and whims are grounded out into shapes and patterns and moulded into a workable design. Now that your abstract sketches and diagrams of the zone plans have been transferred on to a base plan – accurately to scale – the shapes of the spaces can be refined and adapted, the circulation routes can be altered to suit the new firmer shapes, and so the preliminary sketch design starts to emerge. It is a fact of life that pure aesthetic principles will be compromised to a lesser or greater extent by cost constraints, planning matters, restricting briefs, the need to accommodate statutory undertakers, local planning authority criteria, existing site features and restraints, and cautiousness from or compromise with other members of the design team.

Constant cross-reference to the desired moods or context will be necessary to help the designer to choose the right shapes, patterns, colours and materials. The mood will determine the flavour of the design layout, which should be enhanced by the choice of the materials of construction, as the next chapter will demonstrate.

Summary of contextual methodology

The following sequence of work may be of assistance in working through the early part of the design process outlined in this chapter:

1. Set out your key word diagram and adjust your marker point along the scale to where you judge is right for your site's context. Then set out all the client information and site appraisal data on zone plans.
2. Prepare further zone plans for the space required for each activity/function and determine the style of these from the key word diagram. Show the main enclosing elements and structural planting. Show the circulation routes for both pedestrian and vehicular traffic in diagrammatic form.
3. Prepare pencil sketches and notes, both three-dimensional and in plan form, overlaying the zone plans and refining the shapes of the spaces and the junctions and connections between them.
4. Crystallize your first ideas into a coherent design concept, discarding those ideas that do not fit your concept, as governed by the key word diagram. Think again about space creation and spatial hierarchy. Apply pedestrian and vehicular circulation patterns in more solid form, thinking about path hierarchy, widths, surfacing type and junction treatment. Block in the structural planting and define planted areas in more detail, ensuring that such structural elements have either one side fixed to a boundary or face out both sides – each side with a space-defining purpose. Rough sketch plans, perspectives, written notes and above all diagrammatic zone plans will be produced to explore and record the thinking process.
5. Expand on the best ideas, making each part clearer, and start to think about suitable materials, edgings/trims and junctions between surfacings and level changes.

6. Think about possible methods of construction and how these might affect the junctions of paths, surfaces and structures.

7. Draw the sketch ideas to a larger scale (1:100) and fit the design to the site more accurately, adjusting the design where necessary.

8. Determine practicable solutions to service and drainage routes, and surface water drainage falls and cross falls, to design out potential ponding problems. Design, at least in outline form, any special features such as walls, steps, retaining structures, ponds, rockeries, pergolas and so on.

9. Work up and take initial quantities for a preliminary cost plan and, if the proposals are within the available budget, work up the design to a presentation standard for client liaison and approval.

CHAPTER 9

Progressing concepts to final drawings

Refining the sketch design

AS discussed in the previous chapter, the sketch design is the next stage of the design process. This drawing firms up the different zone plan sketches and notes into a more coherent and detailed form (see Figure 9.1). Existing landscape features that are to be retained may well require attention to make good any damage or fault. These features could be hard or soft elements. Some of these items will be of primary or strategic importance or may be protected by preservation orders. Others will be of secondary importance. The designer will determine those items that will be affected by the development proposals and may well influence development layouts to retain as many of the primary or strategic features as possible. Secondary features may be sacrificed while others will be retained where they are not affected by the proposals.

Soft features such as trees and hedges identified as requiring attention may well require works to improve their condition (for example crown raising, thinning, dead wood removal, pollarding, etc.) or treatment for pests. An example of a hard landscape feature would be a brick or stone wall, which may be protected under planning legislation and yet will require works to ensure its structural integrity, which might involve underpinning or re-pointing. Typical soft and hard features to be retained and made good in this way might, for example, include walls, fences, railings, gates, pergolas, summerhouses, hedges, trees, attractive shrub borders, etc. Such items will have been attractive enough to want to incorporate into a new design, which will add maturity to the scheme and demonstrate the designer's ability to integrate the old with the new to the benefit of the scheme. All items that are not to be retained can be removed from the base plan. Such features might include an old rotten timber shed, a rotten tree, a concrete slab, an unsightly or unsafe free-standing wall and so on.

The shapes and framework of the design will be modified to fit the exact site dimensions. Paths and focal points will be adjusted and manoeuvred to suit. Junctions of spaces will begin to be defined, as will changes of level, kerbs, fences, hedges and so on. Paths will be given indicative materials and edging treatments. Textures and symbols will be used to depict materials.

Now that the designer is considering the more detailed elements, there

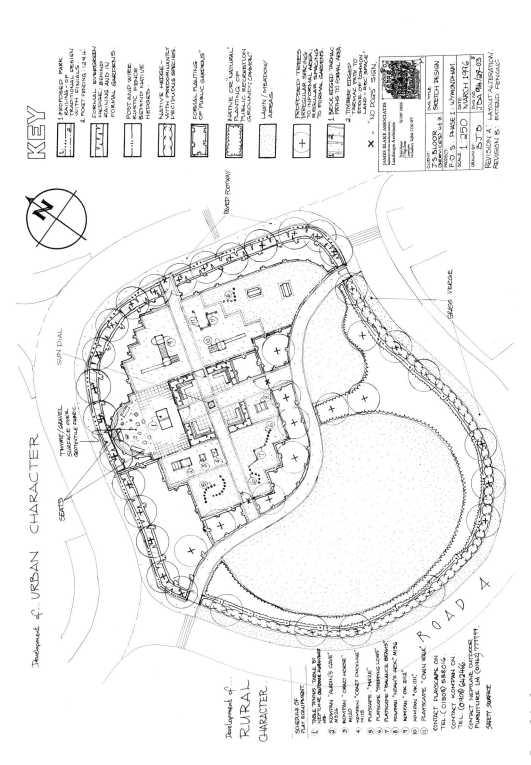

FIGURE 9.1 SKETCH DESIGN

will be further reference back to the site survey data and analysis. Such matters for consideration might include items to be screened, shelter to be provided for, existing trees to be protected, boundary fencing types to be considered for privacy and security reasons and so on. Any planning regulations that affect the site will be taken account of, such as Areas of Outstanding Natural Beauty or Conservation Area status. Tree preservation orders will be considered, which may protect hedges or groups of trees as well as individual trees. Boundary fences proposed may require planning permission if 1 m or higher, adjacent to a highway or in the case of boundaries elsewhere, if 2 m or higher. Site access may have to be improved with new visibility splays so that cars exiting from the site can see oncoming traffic at a reasonable distance. This line of vision may be up to 4.5 m back from the main road and possibly up to 70 m up the road. No tree, shrub, fence or other obstruction over 600 mm high will be permitted within this area.

There may be other planning and legal issues to consider, such as ownership rights over existing hedges and fences. Rights of access, rights of way and easements may be present, authorizing access by an individual or by statutory utility companies. Such companies have policies on how close you can plant to service runs and on which varieties and heights are permitted within a given distance. All such planning and legal factors have to be considered at an early stage in the design process and will modify to a lesser or greater extent the final layout.

Client liaison

The sketch design will be further refined to take into consideration all the site and planning issues. When this is done, the designer will need to show the plan (perhaps for the first time) to the client. Only now will the designer feel that he or she has presented the idea sufficiently well for the client or member of the public to be able to understand the ideas. Considerable effort and mental anguish may have gone into this one outline drawing and the ego may well suffer a knock or two if the client does not like it, but it is essential that the client be approached at this sketch scheme stage to ensure that the client's wishes are satisfactorily interpreted. The designer should carefully work through all the points in the brief with the client, referring to the plan.

The client may well have changed his or her mind by this stage and is, of course, at liberty to do so until a final and approved design is reached (though you may be able to charge extra for additional time expended over such changes). The client will scrutinize the drawing and ask many searching questions. If your client is a local authority then this scrutiny will come from both the council officers and from the public at large who will be the end users. Public consultation is vital when designing with public money, for reasons of political and financial accountability.

Amendments and revisions

Once the comments have been received and matters discussed, amendments to the plan can be made, which will then be marked 'Revision A' (date it and add a short sentence outlining the revision). It is the designer's duty to advise the client of the effect of these changes upon the design and the site but the client may or may not accept this advice. The amended plans will then be represented to the client and further points may be raised. It is important to encourage as much scrutiny as possible (against a natural desire for your design to remain intact) because there are immense problems with changing the scheme after the job has been finally designed, agreed, working drawings and documents prepared, tendered and perhaps even the contract awarded.

Final design drawings

Selling a dream

The final design drawing(s) is the final and most carefully presented drawing to the client (see Figures 9.2 and 9.3). The designer is selling a dream, selling his or her creative ideas. If the client does not like the design, the designer will have wasted a lot of time. The landscape designer has a duty to the client to be precise in communicating ideas and design solutions. It is essential in so doing to set out all information in a recognizable fashion, using the established conventions for graphic communication on drawings. However, there is more room for graphic styling and individuality of pen work with a final design drawing because the designer is trying to convey the feeling and mood of the proposed space. It is worth the designer investing time to draw the plan using the very best graphic skills.

Essentially the final design drawing is the culmination of all earlier sketch drawings and research into the site factors and client requirements, presented for final approval by the client. The graphics should be clear and should explain everything that is proposed so that the client can be as certain as possible that it is going to get what it wants. Subsequent drawings will from now on be for an entirely different purpose. The designer will now concentrate his or her efforts on producing drawings that help a builder to construct the proposed design. These later drawings are called 'working drawings' because unlike the design drawing they are trying to convey just one thing, how the materials of construction fit together and their arrangement. With such drawings there is no room for ornament or graphic styling; they are very dry and follow closely drawing conventions.

Working drawings

Working drawings are for the contractor to enable the designer's scheme to be

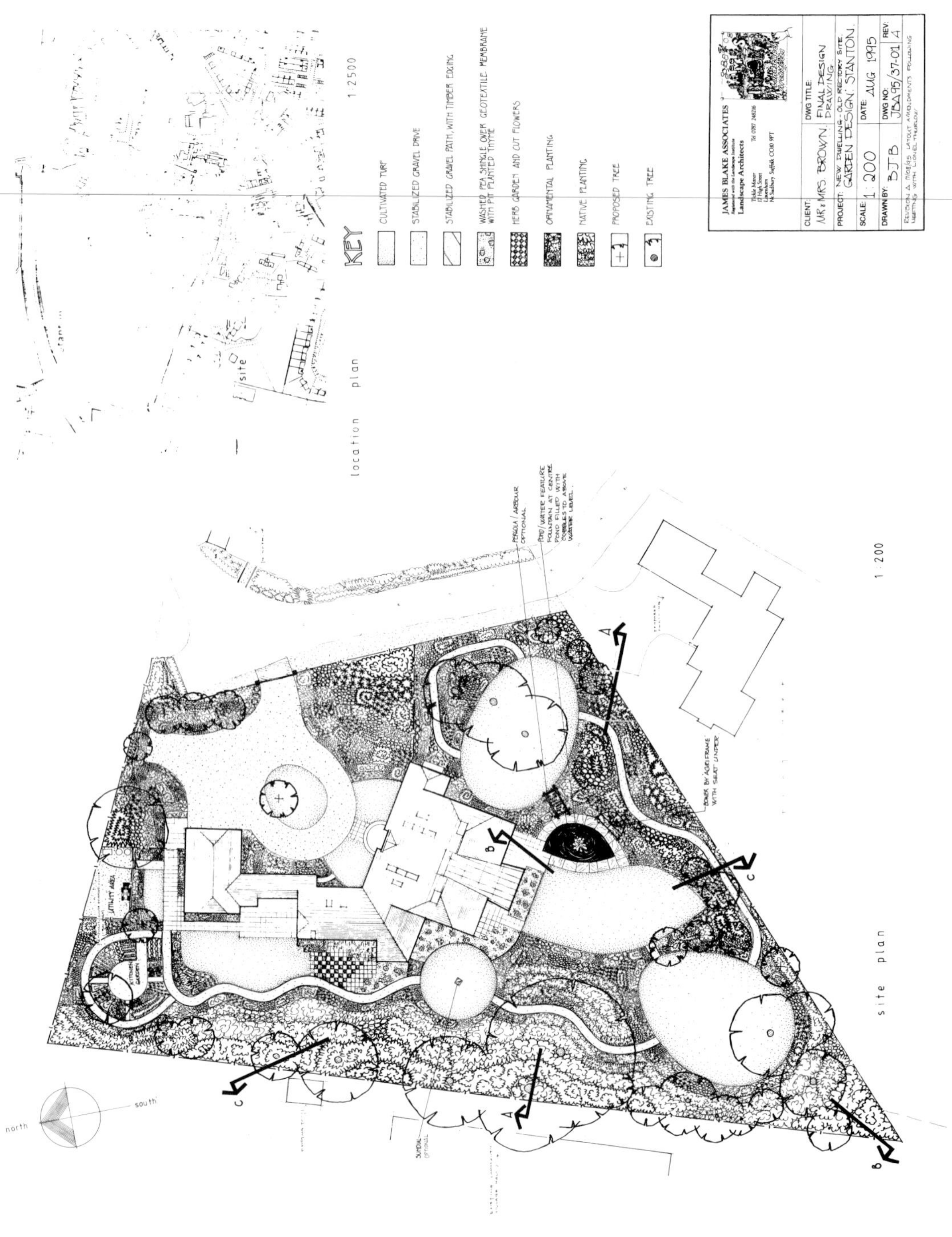

FIGURE 9.2 FINAL DESIGN DRAWING (presentation plan)

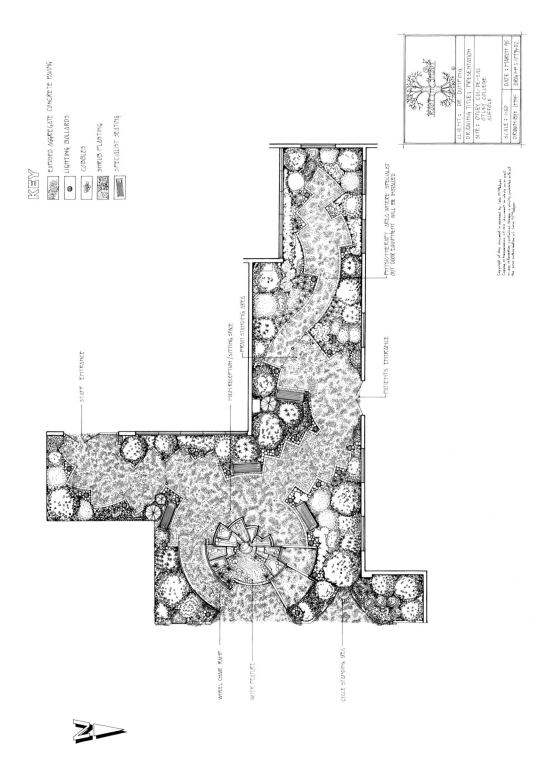

FIGURE 9.3   PRESENTATION PLAN

built. These drawings will be precise and clear outline drawings with the minimum of embellishment to explain simply to a contractor how to build the design. Working drawings either explain the general arrangement of materials or are 'blown up' details of the component building materials, explaining how they fit together. These details will be referred to on the general arrangement plans so that a contractor will be able to find any single piece of information that it requires to build the scheme. For example, if the contractor wanted to fix the rails of a timber fence then the foreperson would check the general layout plan to locate the position of the fence and find the number of the detail explaining how it is constructed. From this detail the size of timber required, the fixing type (screw or nail and its size) and any stain treatment specified can be found.

All such working drawings (but not design drawings) including both general arrangement plans (sometimes layout plans) and construction details will be packaged with supportive specification documents to seek competitive prices (tenders) from contractors. Both the drawings and written documents are therefore known as tender documents. Once a contract has been let to a selected contractor, these same documents will be referred to as contract documents.

Hard works layout plan

This plan is for showing the layout, location and arrangement of all the elements of the plan. It acts as a reference point for all the specialist component pieces of information (see Figure 9.4).

Soft works layout plan (or planting plan)

The planting plan (see Figure 9.5) is a drawing that has three basic elements. First, in the centre of the sheet there will be a plan view of the planting beds. The planting beds are drawn using the British Standard line type (BS 1192: 1984 Part 4) for proposed planting using a 0.4 weight pen (see Chapter 15). The most commonly used approach is called block planting. The planting beds will be divided up using a thinner weight pen than that used for the bed outline, forming blocks or areas within the bed where specific plant varieties (or plant mixes) can then be labelled. The labels involve drawing a pointer line using a thin weight pen (0.18) extending from the block to the side of the plan (or to clear white space on the plan). At one end of the pointer (the centre of a block) will be a large full stop, to mark the end. At the other end of the pointer will be the label (plant quantity and name). This process is repeated until all the blocks within each bed are labelled (with the quantity written before the plant name).

The quantity of each plant variety needed to fill a block will be determined by the area of the block. The area will be multiplied by the number of plants required per square metre. The number of plants needed per square metre varies according to such factors as the growth rate of the plant variety, its size when planted, the immediacy of the impact required by the client and so on. A variation to this 'block planting' method is to subdivide the bed into flow-

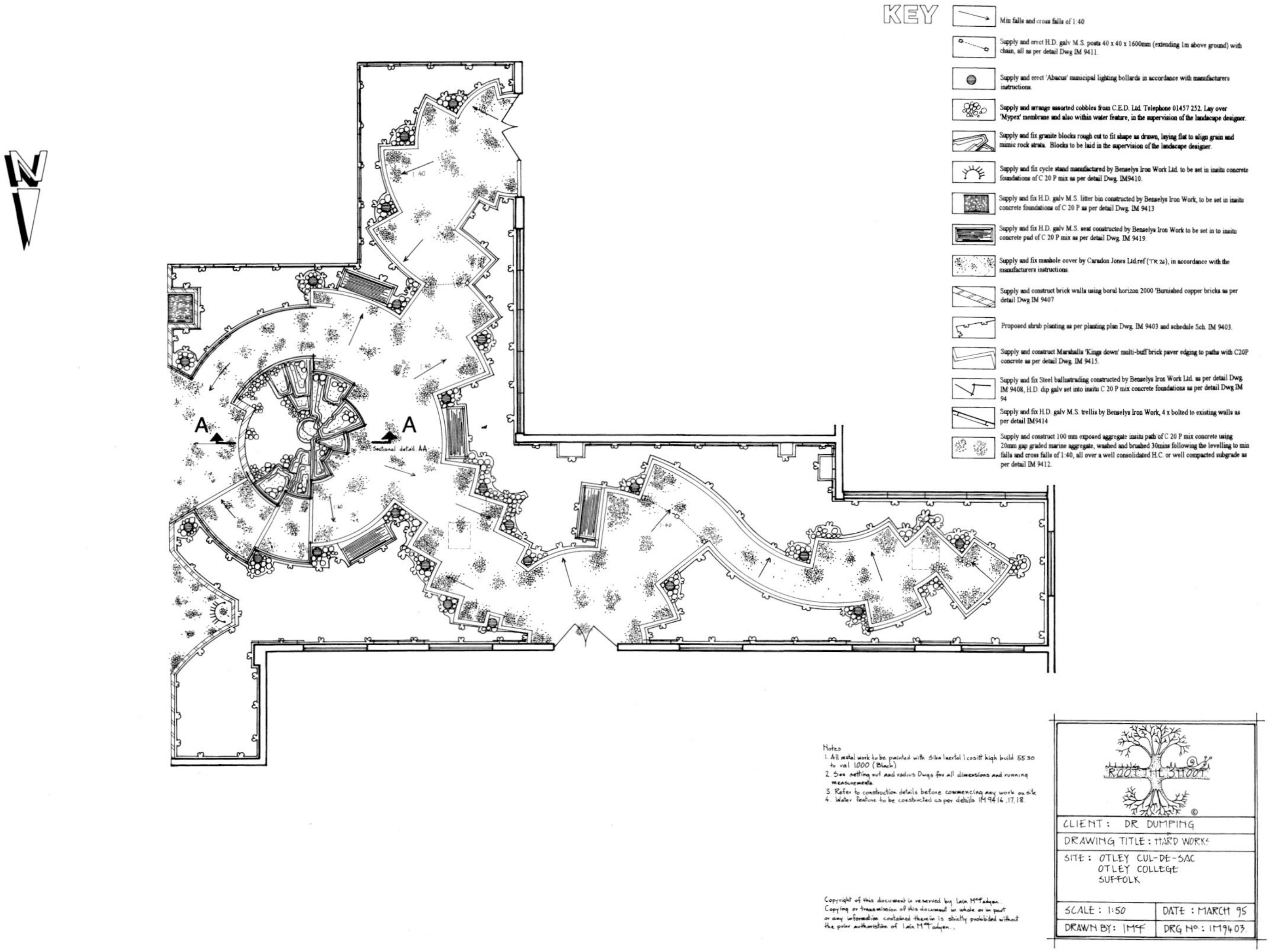

FIGURE 9.4   HARD WORKS LAYOUT PLAN

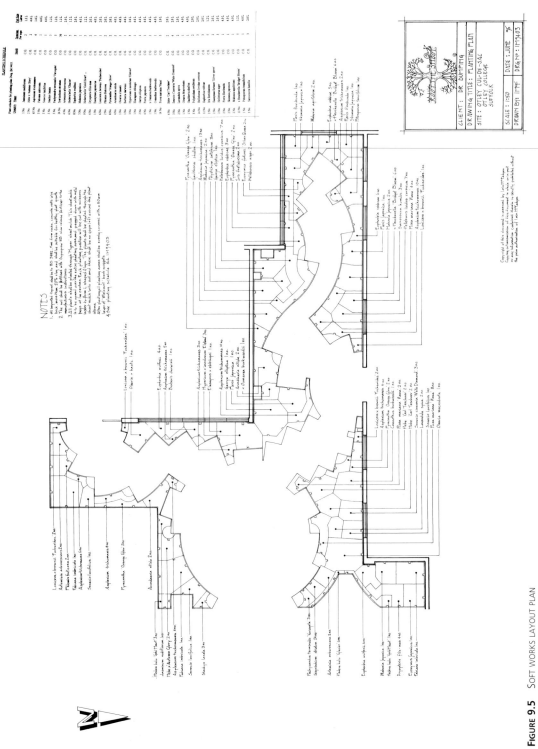

FIGURE 9.5 SOFT WORKS LAYOUT PLAN

ing wedges, supposedly to achieve a more natural effect. Personal experience suggests that block planting provides sufficiently subtle effects and makes it certainly much easier to measure quantities. The flowing method might arguably be more suited to native planting, but it is important to realize that shrubs planted in blocks grow into clumps and lose any regimentation or artificiality almost at once. Indeed it is hard to spot any block formation even immediately after planting. Returning to block planting, the size of the blocks will depend upon the context, native or ornamental. Ornamental beds require smaller blocks than native ones, and a rockery or herbaceous border will require smaller blocks than a structural shrub bed. Similarly, planting along a roadside will be bolder and use much larger blocks than, say, for that by the front door to a house in a suburban cul-de-sac, where the planting will be studied at closer range and at much slower speed.

Second, the planting plan will have a separate list of all the plant varieties selected, written in alphabetical order and set out as a schedule of text, which will include the following information in columns, listed in the order below. The quantity is usually given as the first column, being the sum of all the different blocks in all the beds for each variety. This figure is then followed in the next column by the plant name (sometimes an additional column is given for a plant code which is an abbreviation for the plant name and made up by the designer for use when space does not allow for the whole of the botanic name). Following the plant name will be columns for stock type (container grown, open ground and bare root, containerized, root balled and burlapped, etc.), spacing (the number of plants per square metre) and stock size (the pot size which is measured in litres, and the tolerance allowable for height and spread).

The planting schedule is written in space available surrounding the plan(s) of the planting beds so that it can be referred to easily by the contractor on site without foraging for separate scraps of paper. The main purpose of the planting schedule is to assist the contractors with ordering the plants from a nursery.

Third, the planting plan will provide some specification notes that relate specifically to the site such as details of climber training and support, soil amelioration required, protective measures for existing trees and shrubs, tree staking and ties, etc.

Preparing planting schedules

Great care should always be taken with the preparation of planting schedules, as they are ideally prepared to a standardized format on computer. It is inevitable that old schedules prepared on a computer will be reused and amended for new sites. It is all too easy to alter stock sizes without adjusting spacings or vice versa. Planting schedules are necessary to assist landscape firms with pricing the works. They will send the schedule to nurseries in order to obtain quotations for the supply of the plants. If the information on the planting schedule is incorrect then the plants supplied will be incorrect. Careful choice of the appropriate stock size is essential. While many schemes can happily utilize commonly available nursery stock, that is 2 litre or 3 litre container grown plants (at around 450–600 mm height and spread), there are some situations that will require instant impact. Such schemes will require 10–15 litre container stock (900–1200 mm height and spread). Examples of sites requiring instant impact might include show house complexes for housing developments, family road-side restaurants and hotels.

Stock type must also be selected with care and examples of commonly available stock types include the following:

1. Open ground (field grown in rows – for November to March planting only).
2. Container grown (grown all their lives in a pot or liner).
3. Containerized (grown open ground and then potted up for sale – usually to be avoided).
4. Root balled and burlapped (grown open ground but lifted from the nursery as a root ball which is wrapped in hessian sacking and tied to prevent the roots drying out – for November to March planting only).

The stock type can be selected according to the budget, planting season and availability. Open ground stock is much less expensive than container grown, but must be planted during the dormant season. Container grown stock can be more consistent, but is more expensive than open ground and can, weather permitting, be planted at any time of the year.

For native trees and sub-canopy layer shrubs the stock type category (set out above) is followed by the stock title, which could be any of the following: seedling, transplant, whip, feather, half standard, standard, selected standard, heavy standard, extra heavy standard or semi-mature.

The last columns are devoted to the plant size, which may be given as a pot size (in litres) for container grown plants, height for native trees/shrubs or both height and spread. Even where a pot size is given, it is best to list also the height and spread, as it is common to find recently potted 2 litre shrubs in 3 litre pots being delivered to site and, if these are accepted, a contractor pricing the scheme may have gained a price advantage over other contractors who secured true 3 litre shrubs. Some stock is only available in certain sizes.

Stock size is one of three main factors that determine plant spacing, the others being the plants' growth rate and the desired length of time before a mature, established effect is achieved. Some tree belts might have plants spaced at one plant for every 3 m, while small herbaceous plants might be planted at nine plants per square metre. Spacing of most landscape shrub material will usually be between one plant per square metre and six plants per square metre.

Near to the planting schedule will be placed the specification notes. These can be standard notes such as 'No plant substitutions will be accepted unless first confirmed by the landscape designer', or specific to the particular site such as 'Specimen *Ginko* tree to be planted in specially prepared tree pit – with underground guying – all as Detail No.1'.

Setting out plan

The setting out plan shows just the outlines of the design. These key lines are then located by measurements, using radii, running measurements, offsets and simple dimensions – but always from fixed points such as an existing building, wall or fence being retained. Setting out measurements will be given for all proposed works, both hard and soft. The setting out plan is vital for the contractor to ensure the design is correctly constructed and that proportions and shapes are true to the intention. Failure to pay sufficient attention to setting out information will not only result in a poor interpretation of the design, but may well mean that material quantities are also incorrect.

Demolition plan and dilapidations plan and schedule

A demolition plan will show the whole site and highlight features, structures and surfaces that are intended to be demolished, excavated and carted away from the site (although some material may be retained as hardcore for the construction of sub-bases to proposed paving surfaces). This plan will be diagrammatic showing just the outline of the existing structures and surfaces, which will be hatched and cross-hatched and labelled in a key (legend). These features will either be retained or specified to be demolished or excavated.

Dilapidations are existing items to be retained but which are in less than perfect condition and require improvements. A schedule of dilapidations may be drawn on the plan and will comprise a list of the items to be retained, their grade, condition and works required to make them good. Such plans (which will include tree survey and surgery plans) will require a schedule with columns, starting with the number reference, then the item title, its size, its condition and finally a description of the remedial works required to improve it satisfactorily. Dilapidations can be split into two parts, those for hard landscape elements and those for soft elements. Hard landscape items might include walls that require re-pointing, paving slabs that have settled or cracked, fencing that

---

FIGURE 9.6   SCHEDULE OF DILAPIDATIONS

Ref	Item	Condition	Works required
H1	Brick wall	1.8 m x 10 m bricks sapping	Cut out damaged adj. to garages and re-point.
H2	P.C.C slab paving	8 m slabs uneven, some cracked	Lift and cart away × 1.5 m, to front door; some weed affected slabs; reform base, supply and lay slabs
H3	Palisade	Timber fence, many panels broken	Carefully break out 1.8 × 32 m 14 No. panels in total out to boundary and replace to match

Tree No.	Species	BS 5387 Cat.	DBH (mm)	Age	Height (m)	SULE category	Crown spread (m)	Minimum distance (m)	Visual	Problems	Work required
T1	Fraxinus excelsior/ common ash	C	150	SM	10	2	4 (S) 2 (N)	4	Weak sparse crown	Shaded by adjacent trees – dead wood	Fell to ground level – to avoid competition
T2	Quercus robur/ English oak	A	650	M	18	1	18	10	Prominent skyline tree	Dead wood – ivy	No work required
T3 Tag No. 845	Acer pseudo-platanus/ sycamore	D	300	M	10	4	4 (E) 4 (N) 8 (S)	5	Minor screening value	30° lean into site – dead wood – weak fork at 2 m	Fell to ground level because of weak fork and leaning stem
T4 Tag No. 846	Quercus robur/ English oak	B	350	SM	10	1	12	12	Dominant – prominent skyline tree	Dead wood – previous surgery work	Remove dead wood – reduce crown to increase light to adjacent properties
T5 Tag No. 867	Quercus robur/ English oak	C	300	SM	8	3	14	9	Rounded crown – co-dominant	Bracket fungus on wound – wood boring beetle – previous surgery work – evidence	Fell to ground level because of fungus and evidence of rot
T5A	Quercus robur/ English oak	C	250	SM	7	2	8	9	Group amenity value	Dead wood – evidence of rot	Fell to ground level because of potential rot to provide greater light for adjacent trees – to make way for development

Tree No.	Species	BS 5387 Cat.	DBH (mm)	Age	Height (m)	SULE category	Crown spread (m)	Minimum distance (m)	Visual	Problems	Work required
T6 Tag No. 847	Pinus sylvestris/ Scotch pine	B	158	SM	8	1	6	5	Some value – visible from adjacent field and dwelling	Dead wood	Fell to ground level – to allow light to plot
T7	Pyrus communis/ pear	D	120	OM	5	4	6	N/A	No amenity value	Extensive rot – dead wood	Fell to ground level because of extensive rot
T8 Tag No. 848	Fraxinus excelsior/ common ash	B	160	SM	10	1	8	8	Prominent skyline tree	Ivy	Crown reduction to increase light to adjacent development
T9	Quercus robur/ English oak	C	450	M	12	2	8 (W) 4 (SE)	10	Co-dominant – high amenity value	Dead wood – ivy – previous surgery work – leaning 10° west	Remove dead wood
T10	Pyrus communis/ pear	C	150	M	7	3	5 (S)	6	Low amenity value	Ivy – previous surgery work – shaded by adjacent trees – ash regeneration under-storey	Fell to ground level – increase light to adjacent trees – clear ash regeneration

FIGURE 9.7   A TYPICAL TREE SURVEY

has damaged panels and so on. Figure 9.6 sets out some typical hard works dilapidations.

Soft landscape elements often need improvement works too, and these can be scheduled in much the same way. Such works may include tree surgery, shrub pruning, replacement of dead, sickly or diseased shrubs or of shrubs that have become too big for the scale of the setting. Dwarf conifers will soon destroy the illusion of miniature scale on a rockery if they are allowed to grow too big. Such trees need regular replacement.

Tree surgery plan

A tree surgery plan showing existing trees and vegetation to be removed, retained on the site as existing or retained but modified by surgery may be provided. The trees and other vegetation need to be assessed and graded. The items can then be listed and comments written in a scheduled form suitable for pricing by tree surgeons. A typical survey for surgery works is set out in Figure 9.7.

The grade given is based on the assessment by the inspector as to the health and visual amenity of the tree. The assessment is made against precise criteria set out in BS 5387. In brief summary, grade 'A' means a tree of excellent health, vitality and visual amenity. Such a tree will be prominent and making a major contribution to the landscape. Grade 'B' will be a tree of secondary amenity value, but still being of general good health and condition. Grade 'C' covers trees of satisfactory health and vigour but of lower amenity value. Grade 'D' describes trees that are either dead, dangerous or diseased and dying. These categories are often considered insufficient in themselves as a highly attractive, prominent and mature tree may have a mild infection of honey fungus that will eventually kill the tree. The SULE (Safe Useful Life Expectancy) method of tree assessment categorizes the trees according to their anticipated safe useful life in five bands: grade 1 – over 40 years; grade 2 – 15–40 years; grade 3 – 5–15 years; grade 4 – fell within 5 years; and grade 5 – immature saplings. While the British Standard is currently out of fashion, it is still valuable for distinguishing a tree's strategic importance to the wider landscape setting, in order to assess its importance in planning terms.

These same methods of assessment can be used for assessing shrubs and hedges.

Existing site plan

Sometimes it is useful to annotate the survey drawing and make it a presentable existing site plan to assist the contractor in tendering. This is good public relations – showing off the contrast between before and after.

Construction details

There are drawings to show the detail of how an element, such as some steps, a

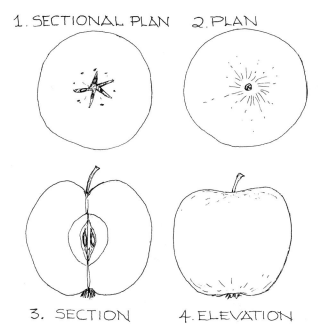

1. SECTIONAL PLAN   2. PLAN

3. SECTION   4. ELEVATION

FIGURE 9.8 SECTIONS AND ELEVATIONS

1. A sectional plan is where the viewer looks down on an object from above, but the view is taken from a point half way down or more the object being viewed. This is the most commonly used plan in landscape design because it avoids obscuring areas of the plan by tree canopies.

2. A plan is a bird's-eye view where the top of an object will be seen. With a landscape such a view would include tree canopies which might obscure crucial elements of the design.

3. A section is a view of a slice through an object end on, so that you can see inside it and in the case of landscapes below ground.

4. An elevation is an end view of an object looking at the outside and in the case of landscapes showing only those features above ground, including items in the distance although drawn in fainter lines.

pergola, fence, wall and so on, is actually constructed. Using plans, sections and elevations and notes, each element can be explained in great detail and at large scales such as 1:10 and 1:20.

Sections

Sections are detailed drawings on plans (although these are sometimes prepared on separate sheets) used to help explain the spatial arrangement of layout plans where the section will traverse the whole site. Where a section is taken across many different materials it is referred to as a running section, and such sections are necessary to show how each material fits together with the next. Sections are essential for construction details, to show the component detail of each element and how the materials are assembled (see Figure 9.8).

Elevations

Elevations are most often used for construction details, to show the appearance of elements that have many different sides and at the design stage to show, for example, the landscape that will appear against building façades.

CHAPTER 10

Cost estimating and tender documentation

Cost estimates

A cost estimate is a breakdown of each part of a design into components (such as shrub planting, turf, gravel paving, brick paving, fencing, tree planting, etc.) which can all be measured. Against these measurements can be put a price and therefore a cost estimate can be prepared. Such a cost estimate will help to ensure that the designer's proposals are not too expensive and that the scheme will not exceed the client's budget. If the design cannot be achieved within the budget available it may need to be scrapped or else phased.

Cost estimates are useful to ensure that the designer makes the correct choice of materials. It may be that the design requires specific materials to make it work (visually or functionally). Compromise on cost grounds itself does not always matter if the second or third choice materials perform or fit the context satisfactorily. Only if they do not perform sufficiently or possess the right aesthetic will the designer require a fundamental design change.

Cost estimates are always useful to awaken the designer to the many parts of the design, in order to assist with the later preparation of bills of quantities. By estimating the cost, the designer will be forced to examine (in itemized detail) the entire scheme. This will help the designer to realize the complexity and sequence of the works and therefore provides the basis for setting out schedules or bills of quantities for optimum clarity. The knock-on effect will be to allow contractors pricing the works (tenderers) to pick up the scheme from cold and understand it quickly, so that they can price it properly in the short time allowed for tendering.

How to set out a cost estimate

Cost estimates need to be clear and concise. They must be well ordered, accurate and complete. Cost estimates are set out as a table, often on ruled paper, and have a standardized format of columns – each with a title. The format is ideal for computer use and can be set up on a spreadsheet or word processor. First, list your items in the correct sequence. Second, 'take off' quantities from the plan – it is essential to be accurate! If in doubt, be cautious and round up figures. Third, research rates. Look up copies of old tenders, ask other professionals, ask a tame contractor, use appropriate reference books, try to work out the processes yourself and calculate the labour required, materials, overheads and profit. You will find that the cost estimate is a good reference or basis from which to produce a full schedule of quantities, saving time and helping to avoid mistakes. The typical format for a cost estimate is set out in Figure 10.1.

FIGURE 10.1   EXAMPLE OF A COST ESTIMATE

Item	Quantity	Rate £	Price £
1. Excavation of tree pits for tree planting – 1 m^3	26 No.	25	650
2. Add back suitable topsoil to the tree pits – 1 m^3	26 No.	20	520
3. Cultivate topsoil and fertilize and prepare beds for planting	575 m^2	1	575
4. Plant trees in tree pits – to include stake and ties and watering in	26 No.	55	1 430
5. Plant shrub beds including sheet and bark mulches	575 m^2	16	9 200
6. Plant climbing shrubs – to include trellis supports	14 No.	35	490
7. Plant climbing shrubs – to include wire support	26 No.	15	390
8. Supply and spread grass seed to lawn areas	909 m^2	0.97	882
9. Maintain all planted and seeded areas for 12 months			1 800

Total estimated cost of works	£15 937.00
Contingency sum	£1 000.00
Total recommended budget	£16 937.00
VAT charged at the standard rate	£2 963.97
Total anticipated cost of landscape works including VAT	£19 900.97

Component parts of a cost estimate

Item

Item is the term meaning each and every component element of the scheme – brick paving, planting, turf and so on are all items. For a cost estimate the title only would be given. There is a correct sequence for these items starting with the demolition of structures and surfaces to be removed, followed by general clearance of vegetation, debris, fly-tipped material and rubbish. Next will come dilapidations (i.e. the making good of defective existing items to be retained), including repairing walls, fences, steps and so on but which may also include soft works such as tree surgery, hedge trimming and so on. Then comes excavation and ground works, followed by hard works, soft works and finally maintenance.

Quantity

The quantity simply means either the area, the length or the volume of each item, as appropriate. For example, planting is measured in square metres (m^2); brick edging is measured in linear metres (m) and topsoil is measured in cubic metres (m^3). Trees, however, are measured by number (No.). A figure will be entered in the quantity column followed by the appropriate measurement symbol.

Rate

The rate is the term given to the price per measurement unit. For cost estimates this will be the price per unit anticipated. That is to say, the rate represents the price that the designer anticipates will be charged by a contractor pricing the scheme. For example, if an area of concrete slab paving were 10 m^2 and you anticipate that a contractor will charge £30 per square metre for that specification, then the rate is £30/m^2. The additional costs of the preliminaries and provisional sums can be added to the rates in the case of cost estimates (to keep them short and simple) but with schedules of quantities these parts will be priced separately. For cost estimates, if the designer has been commissioned only to prepare a feasibility study, then the future design fees can be added to the cost estimate, plus the VAT, which the client may not be able to claim back.

Price

The price is the term for the quantity multiplied by the rate. In the case of a cost estimate it is the total amount you anticipate a contractor will charge for any one item (for the quantity given). All prices estimated should include the anticipated contractor's overheads and profit but not include VAT.

Draft schedules of quantities

Draft schedules of quantities are really just more detailed versions of cost estimates. They are set out in the same sort of format, but there will be much more description about the nature of each item of the works and the items may well be broken down into their component elements. In other words, a cost estimate is a condensed and simplified summary of a schedule of quantities.

Tender documentation

Schedules and bills of quantities

These are a collection of documents that work together to describe the project as a whole; that is, all works proposed and all obligations and factors that affect the carrying out of the works. They are set out in a clear, ordered and familiar way in order to allow fair and accurate pricing by contractors. These schedules together with the drawings are called the tender documents because they are all used to receive tender prices from contractors. The component parts of a schedule or bill of quantities are listed below. The preliminaries, specification, provisional sums and measured works form four separate schedules within the schedule(s) or bill(s) of quantities.

1. Invitation to tender.
2. Cover sheet.
3. Contents page.
4. Instructions to firms tendering.
5. The form of tender.
6. The contract particulars.
7. The preliminaries.
8. Specification (or preamble).
9. Provisional and prime cost sums.
10. Measured works schedule.
11. Collection pages.
12. Summary of tender.

Bills of quantities are a standardized, familiar and precise way to set out all the information needed to ensure that firms tendering are clear and certain of the standard and quality of both workmanship and materials required by the employer. The convention that dictates the format for these documents stems from the Royal Institute of Chartered Surveyors Standard Method of Measurement (S.M.M.) and we have reached the seventh edition – hence S.M.M.7.

The detail of the S.M.M.7 methodology is necessary for building contracts, but because landscape contracts are simpler and of a different nature S.M.M.7 is rarely used in its pure form for landscape design. The landscape profession tends to use a simplified format that does not reduce each element of the works to such fine component detail but instead lumps all the elements for each item together. Documents set out thus are called 'schedules' of quantities.

Component parts of a schedule of quantities

Invitation to tender
The invitation to tender is a circular letter to the contractors that accompanies a set of drawings and the schedules (or bills) of quantities chosen, inviting the contractors to tender for the scheme, giving the tender return date and instructions on how to find the site.

Cover sheet
This sheet of paper will be headed 'Schedule (bill) of quantities for . . . . . [the

site name]'. It will also be dated and at the bottom will be the name and address of the designer, or quantity surveyor if involved.

Contents page

The contents page lists all the parts of the schedule (bill) of quantities and gives their page references. These parts are listed below.

Instructions to firms tendering

This is the first proper section of the schedule of quantities. It includes a description of the scheme and instructions on tendering procedure. Such instructions will include *inter alia* that all tenders must be returned to a set time, date and place or they will be invalidated, that all tenders returned must be in unmarked envelopes, that errors or omissions will be dealt with in accordance with the procedure for correcting errors as set out in the NJCC Code of Procedure for Single-stage Selective Tendering and so on (see pages 205 and 206). A list of all the drawings for the scheme is also included.

The form of tender

This one-page form is for completion by the firm tendering. It is a declaration of the fixed price charged for the works to be carried out in compliance with the contract conditions and should be signed and dated in the space provided by a director of the tendering firm.

The contract particulars

The items that affect the contract and which may vary from contract to contract are presented as a table. They include the commencement and completion dates, the retention percentage, the defects liability period, the maintenance period, the amount of liquidated and ascertained damages, the period of time that the tender should remain open and other such matters. Any changes to the wording of the standard form of building contract are also listed in this section.

The preliminaries

The first of the four main schedules. This schedule (or bill) is concerned with items required in order to operate the contract on the site, for example the means by which the contractor can gain access to the site and the location of the access point. It covers clauses relating to site accommodation, power and water supplies, telephones and temporary site access roads, signboards and hoardings. It covers attendance by site agents and clerks of works, restrictions on working hours and practices such as site burning and so on.

Specification (or preamble)

This schedule (bill) is concerned solely with the quality of workmanship and materials to be used. It is a series of clauses which dictate the minimum standards required of both materials and labour. British standards are frequently referred to and careful descriptions are given for each component element. These elements are set out in a set sequence – approximately in line with the

---

FIGURE 10.2   EXAMPLE OF A SPECIFICATION CLAUSE – TOPSOIL

---

a) Quality of topsoil
Existing topsoil shall be cultivated to depths stated below, and shall then be to BS 3882 1965(78); a good quality medium loam of good heart, free from large stones over 35 mm diameter and all other alien materials, perennial weeds, roots and other plant matter, and not more than slightly stony and slightly acid to neutral reaction (BS classification).

b) Replacement of poor topsoil
The contractor shall assess the condition of the existing topsoil and shall seek the direction of the inspecting landscape architect if there is any soil that is badly compacted, contaminated, stony or otherwise unsuitable for the healthy growth of plants. The contractor shall replace rejected soil with imported topsoil and shall inspect the site before tendering and assess the condition of the soil and include for all such replacement topsoil within the tender price. This will include the removal of unsuitable soil from site. Imported topsoil shall comply with BS 3882 1965(78) and a sample of imported topsoil shall be submitted to the inspecting landscape architect and be approved in writing before imported soil is brought on site. All subsequent importation of topsoil shall be to the same quality and any topsoil brought on to the site without the approval of the landscape architect will be rejected and is to be carted off site at the contractor's expense, unless instructed otherwise in writing.

c) Depth of topsoil
Topsoil is to be evenly and thoroughly cultivated to depths as follows:

1. Proposed grassland – 150 mm.
2. Proposed shrub planting – 350 mm.
3. Tree pits: standard trees – 1000 × 1000 × 750 mm.
                 feathered trees – 550 × 550 × 550 mm.

d) Cultivation of topsoil and final grading
Cultivation of topsoil shall be carried out in suitable weather conditions. No machine having a greater ground pressure than 0.26 kg/cm^2 shall be used and any consolidated wheel tracks shall be forked over to relieve compaction. Final grading is to be carried out to ensure the true specified level and grade, avoiding hollows where water may collect. The use of a heavy roller will not be permitted to remove lumps and any area that becomes unduly compacted shall be loosened by forking.

e) Fertilizer
Apply 100 g/m^2 of Fisons Ficote 140 day 16,10,10 slow release fertilizer to soil strictly in accordance with the manufacturer's instructions.

---

sequence of operations on site – and contain reference to hard and soft materials and workmanship from grass seeding to pavement laying. It is essential to establish your own database of standard clauses and then start all specifications from the entire library of clauses, deleting those that you do not require, so that all clauses that are required are included. This might not be the case if you were tempted to reuse an old specification when short of time. A page of a specification relating to topsoil is set out in Figure 10.2.

Provisional and prime cost sums

This is a schedule (bill) providing for unforeseen or special occurrences.

Provisional sums is a general term covering provisional items, contingencies and day works. Provisional items are sums of money set aside for items that cannot (for various reasons) be quantified at the time of tender. Contingency sums are lumps of money set aside for totally unforeseen items and events that are all too likely to happen and cause extra cost (e.g. buried concrete bases, bombs, etc.). Day works are sums of money set aside for plant, labour and materials for unforeseen work that does not relate to any items measured in the bills.

Prime cost sums are specific amounts of money set aside for specialist pieces of work carried out/supplied by nominated subcontractors or suppliers.

Measured works schedule

The measured works schedule (or bill) sets out in sequence all the items of building works that make up the scheme, with a full description of all parts that will affect the price. This is the main section of the document priced by tenderers. Each page is set out with headed columns as shown in Figure 10.3.

Both a cost estimate and a measured works schedule have the same basic parts. The cost estimate will provide much less detail about the works proposed than will be given in a full measured works schedule. This is because the cost estimate is just a summary of the scheme for the purpose of providing an approximate price of the works. The measured works schedule, however, is

---

FIGURE 10.3   EXAMPLE OF A MEASURED WORKS SCHEDULE

Item	Quantity	Unit	Rate	Price
Planting trees and shrubs				
Shrubs and herbaceous plants Supply and plant shrubs in groups and individual specimens in exact positions as drawn, including 'Mypex' sheet mulch (as specified on plan) and 50 mm of Melcourts 'Rustic Biomulch' and watering in with 20 litres water per m^2, applied with a fine spray hose. As plan JBA00/00-00. Approximate quantity:	502	m^2		
Trees Supply and plant trees, including 1.2 m long stake (0.6 m above ground) 50 mm dia., complete with tie and spacing device set 550 mm above ground; including 300 mm radius of Melcourts 'Rustic Biomulch' mulch around the base of the tree 50 mm deep, recessed 15 mm below the grass level to allow for mowing. Water with 25 litres per tree position on planting. All as drawing no. JBA00/00-00.				
Standard trees: approximate quantity:	17 No.			
Feathered trees: approximate quantity:	4 No.			
Container grown: 35 litre pot:	1 No.			

the section of the larger schedule (or bill) of quantities which will be priced by contracting firms bidding (tendering) for the work. It must therefore be very detailed and thorough. In a measured works schedule a full description is given for each item.

To prepare a measured works schedule (arguably the most important schedule in the schedule of quantities), first list all the items in the correct sequence, then write out everything required to carry out each item, referring to all the component elements. Write out each item as a command, for example 'slab paving: supply and fix 450 × 450 × 50 mm Marshalls "Saxon" buff paving slabs, butt jointed, over 50 mm consolidated depth of sharp sand, over 100 mm of well-consolidated and blinded hardcore, over compacted subgrade, all laid to 1:40 cross falls. Include for a layer of ICI "Terram 1000" to be laid over the consolidated subgrade.' Then, as for the cost estimate, 'take off' quantities from the plan very accurately, being cautious and rounding up your figures. If you have under-measured, then a contractor who spots the mistake will have an unfair tender advantage. This is because the under-measured item can be priced highly, while reducing rates elsewhere, in the knowledge that the designer will be forced to add the extra quantity to the contract on site, boosting the contractor's profit.

In a measured works schedule, the rate will be left blank for the contractor to price during the tender process. There is a total at the bottom of each page of all the prices (not the rates) of items on that page. This page total is carried over to a collection page.

The collection page

A collection page is a single sheet listing all the totals at the bottom of each page (these totals are called 'page collections') and then all the page collections are added up to give the grand total cost estimate or the measured works grand total cost. The measured works schedule (or bill) is one of four schedules (or bills) which put together make up a schedule (or bill) of quantities. Each of the four schedules (or bills) has a collection page at the end. All the four collection pages are then added up on a final sheet called a summary sheet.

The summary sheet

The summary sheet is a page at the end of a schedule (or bill) of quantities that lists the totals of all the four collection pages and adds them up to give a grand total cost.

CHAPTER 11

Tendering the contract documents

Tendering procedure

THE drawings and specifications or alternatively the drawings and schedules (or bills) of quantities are collectively called the tender documents. They should have been prepared in a way that ensures that the scheme is tendered on a fair and equal basis, so like can be compared with like, in order to ensure optimum value for money. If all tendering firms are pricing for the same quantity and quality of both workmanship and materials, then the lowest price must (in almost all cases) provide the best value for money. The only exceptions are when a tendering firm has made many errors in pricing or else there is insufficient control in the documentation over how long the contract will take, and a contractor is able to provide a slow, stop-go service. The independence of the designer can allow 'policing' of the scheme, to ensure that the specified items are carried out satisfactorily to specification. With a clearly defined scheme, fair tendering and then close inspection of the works, tendering almost guarantees that the lowest price is the most competitive price and the best value for money, especially if the firms invited to tender are all of approved competence.

It is essential that the contractors that are put on the tender list should not only be approved as being of general competence but be able to undertake the specific nature of the work proposed. Otherwise the designer may have to spend a large amount of time on site checking and inspecting, and sometimes actually showing the contractor how the work should be carried out. This will be expensive for the designer and for the client. Some firms, though competent, can only handle soft works of small contracts. The client may well turn to the designer and ask why an incompetent contractor was put on the tender list, which could both be embarrassing and lead to a breakdown in relations between the client and the designer, especially if the client is receiving large site-attendance fee invoices.

By drawing on the approved lists of local authorities and landscape practices, firms can be checked out thoroughly to ensure that they are competent. The time taken over the selection of contractors for a tender list is more than saved later when the successful tenderer becomes the contractor on site. The tender process involves the following steps:

1. Assemble documentation.
2. Choose a method of tendering.
3. Choose your tender list.
4. The pre-tender enquiry.
5. Send out the tenders with a letter of invitation.
6. Opening tenders.
7. Examining the tenders.
8. Reporting to the client.
9. Notifying the contractors.
10. The signing of the contract.

Assemble documentation

Print enough copies of the drawings and any attendant loose schedules, and sufficient copies of the bills of quantities for each contractor to receive two copies. Include a prepaid return envelope, addressed and marked only with the words 'Urgent Tender' to avoid any marks on the contractors' own envelopes which might identify the firm tendering. You will need five or six firms for most tender situations but you can use three or four for small schemes. Do not seal the documents until the steps in the succeeding paragraphs have been carried out.

Choose a method of tendering

There are many types of tendering – open tendering, single-stage selective tendering, two-stage selective tendering and so on. The most commonly used is single-stage selective tendering, which is regulated by the National Joint Consultative Committees Code of Procedure, known as the NJCC Code of Procedure for Single-stage Selective Tendering.

The problems with other types of tendering normally preclude their use. Open tendering means that the scheme is advertised and any firm can bid to do the job. With such an approach it is impossible to be sure that a particular firm tendering is competent and indeed you will certainly get a complete spectrum of firms bidding, from the completely inexperienced and incompetent to the highly experienced and competent. To choose the best quality firm for a small front lawn turfing contract would be using a 'sledgehammer to crack a nut', and to choose a one-man band complete with shovel and wheelbarrow would be to court disaster if a wide range of hard and soft works was required.

Two-stage selective tendering is a method where firms are invited by open tender through an advert that usually states the standards required and requests a preliminary package. From this package five or six firms are selected to submit a more detailed tender package. This system ensures that there is a qualitative stage of selection before the competitive element is added. This is a common method of seeking consultants for large schemes. It is most likely that for normal landscape construction schemes you will opt to use the single-stage selective tendering method.

Choose your tender list

It is important to choose the firms to tender carefully to ensure that you get the desired result. They should be firms with a proven track record, they should be able to undertake the size of job tendered, they should be financially sound and they must be of a proven standard equal to the tasks demanded of them by the contract. They should be used to completing tenders and able to hand them in on time. If you do not yet have your own list of such firms, then you may use the approved list from your local authority or from other professionals. You should always seek references from firms that you have not used before and preferably go to see examples of their work. Even this preparation will not guarantee that the firm will be competent and reliable, particularly for clearing up 'snags' at the end of the contract.

Remember that firms that are good one year may have a change of management and be terrible the following year. A good ear to the grapevine and a continual and regular review of the approved list is required.

The pre-tender enquiry

It is essential to telephone and write to your selected contractors to ascertain if they are interested in tendering. This will save much wasted time and expense and, moreover, ensure that you receive sufficient tenders to give you a competitive price. If you make sure that you receive written replies to this enquiry, it will be harder for contractors to let you down later. Some still do anyway, which suggests a shortage of good firms.

Letter of invitation to tender

This circular letter is sent to all the firms that you wish to invite to submit tenders. It will refer to the documents enclosed in the envelope, an outline of the scheme and any third parties to whom the tenderers should refer. The letter will advise the tenderer of the last time, date and place that tenders will be accepted and stipulate that late tenders will be invalid. The letter will state that no qualifications will be acceptable and that errors and omissions will be corrected in accordance with the NJCC Code of Procedure for Single-stage Selective Tendering – either Alternative 1 or 2. The letter will go on to describe how to reach the site. Once this letter has been added to the documents, they can be posted off to the various firms, ensuring that they have at least 28 days to submit the tenders. Less time than this is likely to result in some or all of the tenders being returned unpriced.

Opening tenders

The unmarked envelopes should all be received by a due time and date (clearly indicated on the invitation to tender letter). Any late tenders should be set aside and marked 'Late and Void'. The tenders should then be opened (sometimes in the presence of the client) and examined by the designer.

Examining the tenders

The tender prices should be set out on a summary sheet against their respective company names and any general comments such as how they compared to the cost estimate. Then the detailed examination commences. The tenders should be checked for any qualifications, arithmetical errors and omissions. Then a brief report should be prepared comparing the prices of the different firms, commenting on discrepancies. Any qualifications (such as only being able to hold to the price quoted if the contract period can be extended by two weeks or if plant sizes can be altered and so on) will be noted and in the event of that firm being lowest, then the firm will be contacted and asked to withdraw the qualification or else to withdraw their tender. If the firm withdraws the tender, then the second lowest will be contacted. If this firm has no qualifications, errors or omissions, then this firm will be appointed contractor, and contracts signed. However, if there are any errors or omissions in the pricing, these will be dealt with using the NJCC code. Alternative 1 gives two options: to stand by the tender price or withdraw. This is true whether the price is in error up or down. Alternative 2, however, allows the tendering firm either to confirm their price or to correct genuine errors. If the firm elects to correct the error and it is no longer the lowest tender, then the firm now lowest should be contacted, and if there are no errors etc. then this firm should be awarded the contract.

Reporting to the client

The examination of the tenders and the comparison of them should be set out in a concise report to the client. So long as the firm has no qualifications, errors and omissions (or else they have been dealt with in accordance with the above procedures) and the prices look reasonable compared to the rest, then this firm should be recommended for acceptance. If the prices are drastically different from the majority, it is possible that there is a mistake, or that the firm is trying to buy work to maintain cash flow because it is in financial trouble. If it is just one or two prices that are low then it is likely to be a mistake (perhaps they priced for materials only) and then the NJCC code should be followed. If all the prices are significantly lower then the firm could be in trouble and it may be worth recommending the second lowest tenderer for the contract. It is important to note, however, that in all but this situation the designer should recommend the lowest tenderer, otherwise what was the point of putting them on the tender list? Moreover, it is a complete waste of time for the contractor, who may have spent many hours pricing the tender documents, if the tender is going to be rejected. At the end of the day, however, the tender report presents a *recommendation* for the client who is invited to make the final decision.

Notifying the contractors

It is both convention, courtesy and part of the NJCC code to inform the unsuccessful tenderers of their plight and provide them with sufficient information to assist them with future pricing. This information is imparted by listing the firms invited to tender in alphabetical order and the prices received in price order, separately and below. This should be done only after all tendering procedures have ceased and the successful tenderer has received a letter of intent from the client.

The letter of intent will inform the successful tenderer that it has been selected to become the contractor for the scheme. The designer will then arrange a pre-contract meeting, when the contract (JCLI or other) will be signed by both the client (now called the employer) and the lowest tenderer (now called the contractor). The tender documents are from now on referred to as the contract documents.

Signing the contract

The contractor and the employer sign two copies of the standard form of building contract chosen (usually the JCLI), one for possession by each party. This usually takes place at the pre-contract meeting (an important initial meeting between employer, contractor and designer). The designer and the quantity surveyor (if any) are also identified on the contract cover as agents of the employer. They have a duty to act fairly in the administration of the contract between the parties. However, since the legal case *Sutcliffe* v. *Thackrah* (1974), the designer is not a quasi-arbitrator and, in being fair, should be careful not to expose the client to any risk in certifying payments to the contractor.

Compulsory competitive tendering

It is not only landscape contractors who are asked to submit tenders. Landscape designers may also be asked to submit competitive fee tenders for individual commissions or indeed for much larger contracts involving, for example, the entire landscape service for local authorities. Sometimes these competitive fee tenders are purely concerned with price while at other times they will have a qualitative element, requiring speculative plans and cost estimates as well as supporting documents and brochures. Compulsory competitive tendering or CCT is the term given to recent legislation that demands that local authorities tender all of the work for each of their technical services (including landscape design) to suitable private practices, and also invite a tender from their existing landscape design team (often re-organized to form their own separate practice).

The quality of services to be provided is as important, if not more so, than the fee to be charged, particularly since the fee is a relatively small proportion of the total lifetime building cost. Good quality design and construction can effect a saving in the building cost, operation and maintenance more than equal to the amount of the fee, and there must be a careful balancing of any short-term benefits from reduced fees with the longer-term value for money.

In order that both objective and subjective value assessments may be made it is imperative that the key criteria the client wishes to evaluate are made known at the start of the selection process, or practices will waste time concentrating upon the wrong information and the client may receive a false impression about the consultants and choose one less suitable.

Value assessment is a formalized method of balancing quality and price with the aim of deciding which proposal provides the best value for money. The process of value assessment can be audited and be seen to be open and free from favouritism, influence or inconsistency with public interests. It guards against excessively high or low tender costs and achieves the goal of quality tempered by price. The process should take place in two stages: first, the preparation of a shortlist from respondents to public advertisement and, second, the evaluation of fee bids received.

In preparing the tender documents, the client and professional advisers should first decide on the relative weighting of quality against price for the commission. Landscape design tenders will be judged according to the first stage in the evaluation of prospective tenderers and the compilation of a shortlist. The shortlist will be compiled by analysing the prospective tenderers' suitability for the project in question by way of the following criteria:

1. Eligibility – evidence of required range of professional expertise for multi-disciplinary services. Evidence of professional indemnity insurance. Any joint venture commissions proposed for the venture and a statement of absence of pertinent convictions.
2. Financial standing – audited or certified accounts for the previous three years.
3. Technical capacity – names, qualification and specialist experience available relevant to the proposed works. Methods and procedures for design and project management (CAD and draughting, planning, estimating, cost control, management and supporting consultants, construction contract management) and details of quality management systems. Details of relevant experience during the past three years, supporting services provided by way of reference. Details of staff training and details of the practice's environmental policy. Details of the existing client base and associated workload. Projected workload for staff intended for the commission if successful. Ability to provide quick response. Archiving, information, recall and database procedures compatible with the client's own systems.

The panel may wish to conduct detailed interviews with or request presentation from practices that are being considered for invitation to tender.

The number of consultants invited to submit tenders should not be less than five or more than ten.

The quality criteria will be set out on a tender assessment sheet that details their associated adjusted weightings and the adjusted weighting of price. Each completed tender will be individually judged by each member of the tender panel. The panel may discuss together their individual markings to arrive at a consensus assessment or take the numerical average of the respective markings as the final assessment.

It is recommended that the marking of prices is carried out after the assessment of quality criteria and that the members of the tender panel should be unaware of the prices submitted when exercising their quality judgement.

If successful a firm should be invited to sign an agreement which should embody the brief, the schedule of services, the fee basis and the conditions of engagement. All tenderers should be notified of the decision in writing.

CHAPTER 12

Contracts and contract preparation

The client and the designer

Memorandum of agreement

A memorandum of agreement is a standard form that sets out the names, addresses and telephone numbers of the parties, a description and location of the works and the services and fees that have been agreed (see Figure 12.1).

Clearly, a more watertight agreement is preferable but less frequently used than many would admit to. This is partly explained by the fact that many clients are repeat clients, and a position of trust and understanding has already been reached. In such circumstances, especially where a standard fee arrangement has been negotiated, a client may well instruct works without asking for a fixed fee quotation, and the contract is formed by just two letters, their letter instructing you to do the work (to your standard fee charges) and your confirmation of the instruction.

Works order

Sometimes, larger companies will provide you with a works order and possibly a self-billing invoice too. This sets out the contract terms in the same way as a memorandum of agreement, but comes from the client rather than from the designer. Again this is especially common where a designer's services are retained or used regularly.

The client and the contractor

The main point about the contract and the relationship between client and contractor is that there are few areas of greater potential hazard and litigation in contracts between parties than between a contractor and its employer (the client). When large amounts of the client's money and land are tied up under the contractor's control for a period of time, clearly there can be much cause for friction. If the contractor, the employer or the designer fail to honour or fulfil

FIGURE 12.1
MEMORANDUM OF
AGREEMENT FORM

**SPECIMEN FORM OF
MEMORANDUM OF AGREEMENT**   **APPENDIX II**

between Client and Landscape Consultant for use with the Landscape
Consultant's Appointment.

This Agreement

is made on the   .... day of .... 19 ....

between   ....
(insert name of Client)

of   ....

....
(hereinafter called the 'Client')

and   ....
(insert name of Landscape Consultant or firm of
Landscape Consultants)

of   ....

....
(hereinafter called the Landscape Consultant)

NOW IT IS HEREBY AGREED

that upon the conditions of the Landscape Consultant's Appointment
(................Revision) a copy of which is attached hereto

save as excepted or varied by the parties hereto in the attached Schedule of
Services and Fees, hereinafter called the 'Schedule',

and subject to any special conditions set out or referred to in the Schedule:

1.   The Landscape Consultant will perform for the Client the services listed in the
Schedule in respect of

....
(insert general description of project)

at ....
(insert location of project)

2.   the Client will pay the Landscape Consultant on demand for the services, fees
and expenses indicated in the schedule;

3.   other consultants will be appointed as indicated in the Schedule;

4.   site staff will be appointed as indicated in the Schedule;

5.   any difference or dispute arising out of this Agreement shall be referable to
arbitration

AS WITNESS the hands of the parties the day and year first above written

Signatures: Client .... Landscape Consultant ....

Witnesses:

Name .... Name ....

Address .... Address ....

Description .... Description ....

their contractual commitments or make errors and omissions, then the consequences for each party are grave.

Use of the established conventions for setting out contractual terms ensures that the contract will be more easily accessible and understandable by the client and by other related professionals involved with the scheme, and that there is less scope for ambiguity.

It may well be that if the contract is just for laying a small area of garden turf or even a small garden layout, for example, the designer will not recommend a formal contract but will rely instead on drawings and specifications and an exchange of letters between the client and the contractor. However, for any larger or more complicated site, in order to manage such a volatile and potentially dangerous situation, the contract between the contractor and the employer needs to be clear, precise, familiar and thorough. Clearly, a standard form of contract is required, the basic form of which is known as the standard form of building contract. The construction industry is diverse and therefore there are many variations of this standard form. There is no need to describe these contracts in great detail here as other books have concentrated on this area sufficiently, but the following examples are described to acquaint the reader with the most common forms.

Standard forms of contract

The JCT 1980

It may not be necessary to know every detail of the larger contracts such as the JCT 1980 (Joint Contracts Tribunal 1980), which some landscape designers may rarely come across. This form of contract is purely for the larger construction projects and there are many versions that differ from the original, for example those used by local authorities or the private sector, situations with quantities, without quantities, with approximate quantities and so on.

Local authorities frequently add clauses or modify others so that the contracts fit in with their standing orders and regulations. Such alterations should be kept to a minimum, however, because one of the reasons for using a standard form of agreement is the familiarity shared across the entire building world.

The IFC 1984

The Intermediate Form of Contract 1984 (IFC 1984) is a shorter contract used for medium-sized building contracts. There is special provision in this contract to allow the architect to assemble a list of three approved firms for tendered selection by the main contractor for specialist subcontractor works. When there are landscape works associated with a building construction scheme, the named list of three firms is commonly used and the successful tenderer is appointed as a domestic landscape subcontractor. This system ensures that the main contractor is responsible for all co-ordination and management of the subcontractor. However, there are often problems with late or non-payment to such subcontractors, and tenderers faced with this possibility build in a contingency.

Unlike the JCT 1980 contract, there is no provision in the IFC 1984 for intervening in the event of such problems. Also, the requirement in landscape schemes for the landscape contractor to be responsible, for example, for replacing defective trees for two to three years may well be in conflict with the defects liability period for the main building works, normally just six months. Under this contract the final account cannot be settled until all obligations under the contract have been met and yet the builder may have large sums of money tied up in retentions which cannot be claimed until the replacement period has expired. Sometimes such a replacement period is forced to be curtailed because of this and then the employer has to pay for defective plant stock that dies after the six month practical completion period.

The JCT minor works contract

For small building projects, extensions and single dwellings, a much shorter and less cumbersome contract is needed. For such projects, the JCT minor works contract is used.

The JCLI form of contract

Landscape projects are generally relatively small and uncomplicated compared to building projects and thus a small contract is required. For this reason (and to provide the required familiarity) the JCLI form of contract was prepared by the Joint Council for Landscape Industries using the JCT minor works contract as a basis. The difference between these two contracts is not great, but the JCLI form is tailored to landscape works, possessing additional clauses concerning the liability for defective plants of various types, partial possession by the employer and the liability for plants that have been subject to malicious damage and theft prior to practical completion. The landscape designer should be entirely familiar with the detail of the JCLI form of contract and separate study of this document is essential for efficient and confident contract administration.

A full and lucid explanation of all the above contracts and others can be found in Hugh Clamp's *Spon's Landscape Contract Manual* published in 1995 by E & F. N. Spon Ltd.

Subcontracts

As with the main contracts, subcontracts need to be clear, precise, familiar and thorough, and in addition they must dovetail with the main contract. For this reason subcontracts should be standard form contracts, prepared to relate to the specific main contract used.

When the landscape works are the last phase of a building contract, it is often simplest and most ideal that a landscape contractor be appointed directly by the employer after the building contractor has left site. This avoids the need for two contractors with often conflicting disciplines to be working on the same site, which is fraught with problems. However, there are occasions when the employer requires the landscape work to be completed at the same time as the building work is completed, to help sales (as with new housing or fast food retail outlets) or for security reasons, to ensure the site is complete, established and less vulnerable to vandalism and theft. In this situation the most common procedure is for the main building contractor to employ a landscape contractor directly as a domestic subcontractor. In order to ensure that you get a competent landscape contractor it is preferable that you supply the main contractor with a list of three or more names from your own approved list (or someone else's if you are inexperienced) so that whichever one the main contractor selects, you can be sure that they are capable of doing good work. This is not an automatic process under the JCT 1980, and unless your specification imposes any restriction, the main contractor may add their own pet contractor to the list, who may not be competent. The IFC 1984 contract has provision for such a named list of subcontractors and avoids such problems.

The IFC 1984 contract has a form (NAM T) for the main contractor to use for tendering, and a form (NAM SC) for the firm selected. Unfortunately, there are often problems with builders and their domestic landscape subcontractors, as some firms will try to find fault in order to avoid paying subcontractors. They are able to win tenders if they do not expect to have to pay their subcontractors and indeed some will get the majority of the work completed, sack the subcontractor and pay for another just to finish off the work, fabricating and embellishing any fault that they can find, being optimistic that many small subcontractors cannot afford to take the firm to court. This is especially common in times of recession. It is easy for the landscape designer to become caught up in this process, being asked by both parties to support the other's case, often with further pressure from the architect who may be inclined to side with the contractor – especially if not well-versed in the technicalities of landscape works.

The domestic subcontract arrangement can be an advantage to the designer in that the main contractor takes responsibility for achieving the desired result. In reality, where an architect is running the main contract and the landscape designer is only employed for a limited number of site inspections on the work of the landscape subcontractor, the control over the contract administration is much weaker than when a landscape designer is in direct charge of a separate landscape contract. This is because all correspondence has to be directed through the main contractor, via the architect (contract administrator). This can lead to a plethora of problems, many of which centre around communications and the contractual relationship. For example, near the end of the contract there will be pressure from the main building contractor to complete. Landscape items such as grass seeding (which always happens last of all and is often delayed due to seasonal factors) might be scheduled as a 'snag'. Practical completion would be awarded to the main contractor, releasing half the retention money, and this would be passed on to the landscape contractor. If the landscape contractor were slow to get on site, then became very busy, or indeed if the seeding were to fail, it might be very hard to get the landscape contractor back to site, for the financial incentive would not be there as the contractor would have had the majority of the money.

For these reasons, a separate landscape contract after the builder has left site is always preferable. If this is not possible it may be worth considering the alternative method of a nominated subcontract agreement. There is provision for such an agreement under the JCT 1980 but not with the IFC 1984.

Nominated subcontract agreement

By using a nominated subcontract agreement you have greater control and are responsible for the performance of the subcontractor but the main contractor is also obliged to work with (and make provision for) the subcontractor. There is therefore a risk that the main contractor will claim against the client for any delay caused by the nominated subcontractor (since it is the client's agent – the designer – who has specifically requested this one contractor). Moreover, the main contractor can avoid paying liquidated damages for delays in the works that might be (or might be claimed to be) due to the poor performance of the subcontractor. Liquidated and ascertained damages is the term given to a sum that becomes payable by the contractor to the client when the contractor is unnecessarily delayed beyond the agreed completion date of the works. Such damages are not a penalty but are calculated to reflect the consequential loss suffered by the client. With nominated subcontract agreements (unlike that of domestic subcontractors) the main contractor is not, of course, responsible for errors and omissions made by the nominated subcontractor but, obviously, for this reason any designer must be careful to nominate only firms who are known and trusted to do competent work. The main contractor is entitled to a $2\frac{1}{2}$ per cent discount on the nominated subcontractor's price for payment within 30 days, in addition to any attendance sum added at tender stage.

The main disadvantage of nominated subcontract agreements is often the sheer amount of complicated paperwork required. The form of tender, agreements and nominated subcontract are contained in forms NSC/T, NSC/A, NSC/C, NSC/W and NSC/N. Where the landscape contract is an individual main contract the JCLI is the recommended form, being tailor-made for the landscape industry. Often a main landscape contractor will require subcontractors for major earth moving, demolition, metalwork and so on and for these situations there is a standard form to dovetail with the main JCLI contract.

In summary, while small contracts for garden schemes may have a contract comprising an exchange of letters, perhaps referring to drawings and plant schedules (where such exist), with any larger contract a more formal approach is required. After all, there are so many things that can go wrong during the course of such contracts and these problems can easily lead to disputes between the client (employer) and the contractor, and to expensive litigation.

The JCLI form of contract is tailor-made for the landscape industry and it is for this reason that the JCLI is the first choice for landscape designers.

CDM regulations and landscape contracts

The term CDM stands for Construction (Design and Management) and the CDM regulations encompass the latest EU Health and Safety initiative and cover all building construction work, with some exceptions (small works and entirely soft landscape works). The designer must be aware of the current legislation and a summary of how these regulations will affect the landscape designer is set out below.

The JCLI standard form of agreement for landscape works has been revised to take into account the CDM regulations. There are two main options within this contract as follows:

Alternative A

This form is for projects where the CDM regulations do not apply because none of the work is 'construction work' as described by the regulations. Such non-construction work includes topsoiling, grading, amelioration, planting, turfing and seeding, agricultural and rabbit proof fencing, soft landscape maintenance, and forestry and tree works including all associated excavation and clearance operations.

Even where soft works follow on from civil engineering and construction work, so long as such soft works are carried out as a separate contract after practical completion of the building or civil engineering works, the CDM regulations again do not apply. However, typically the borderline between earthworks (which is construed as construction work) and topsoiling (which is not construed as construction work) is unclear.

The CDM regulations also do not apply when the local authority has the role of health and safety enforcing authority. This is never the case for external works but may well occur when dealing with interior landscape works (that is, if hard works are also involved).

Alternative B

This form is for projects where CDM regulations do apply because the works involve 'construction works' as defined by the regulations. Construction work is deemed to include the following:

- Earthworks
- Hard landscape works
- Drainage
- Demolition and dismantling

- Temporary works
- Any associated works such as clearance and excavation.

Alternative B has three sub-sections, one of which will be applicable to each contract. However, all three of these alternative sub-sections mean that CDM Regulation 13 (Requirements Of The Designer) applies.

1. Alternative B(1) – all the regulations apply because the works fall within the definition of construction works, but alternatives B(2) and B(3) are not applicable.
2. Alternative B(2) – Regulations 4, 6, 8–12 and 14–19 do not apply. This is for private gardens unless Regulation 5 applies (where a householder enters into an arrangement with a developer). For such private gardens the planning supervisor, principal contractor, health and safety plan and health and safety file are not required, but the contractor has to notify the Health and Safety Executive (HSE) if the project involves more than 30 days, or 500 person days, of 'construction work'.
3. Alternative B(3) – Regulations 4–12 and 14–19 do not apply. This is where such projects are not notifiable to the HSE because the construction work will take less than 30 days or 500 person days and there is no demolition or dismantling. Also where there are less than five persons on site involved with 'construction work' at any one time. A planning supervisor, principal contractor, health and safety plan and health and safety file are not required for these projects.

If the landscape works are part of an overall building contract and the landscape contractor is a domestic subcontractor, then all regulations apply to all works.

Design and build contracts

The conventional system of tendering schemes prepared by an independent designer to an approved list of contractors has in recent years increasingly fallen from favour. As an alternative, client bodies have begun actively to consider 'design and build' companies instead. This move has been partly due to 'trend' but also to reasoning of the different advantages and disadvantages of this type of contract – and indeed some arguments for the 'design and build' approach do have substance. Design time is saved because the designer can be far less pedantic about specifying every last nut and bolt and preparing watertight tender documents. Great attention to small details is normally vital in order to prevent contractors from using the cheapest and least durable option available so that they can gain a tender price advantage over the more conscientious firm, which is pricing for the most appropriate option. With design and build there is less emphasis on price competition in favour of overall performance, and therefore the contractor can build in a rate that covers the use of a suitably durable material.

Like every coin, there are two sides to this situation. It may well be that where fixings (for example) are not specified by the designer, the contractor will indeed use a suitable fixing. However, it is quite possible that the contractor would choose one that is not appropriate and, though the contractor may be liable for this error, the client then suffers the problems of the item failing and the inconvenience of arranging for someone to put it right. At least with the tendering method every detail (including fixings) can be carefully specified and there is perhaps less risk for the client.

The designer–client–contractor relationship

As explained in Chapter 5, for design and build projects the client approaches a building firm directly, and this firm will either employ an outside designer with whom it has built up a close working relationship or else will have a designer in-house on the pay roll. The client will have selected the firm for one of the following reasons:

1. The client is a layperson and does not know the value of using an independent designer and tendering the works, and may pick a contractor out of the Yellow Pages.
2. The client chooses a known design and build firm because of the reputation of the designers in-house or as a result of seeing some examples of their work.
3. The client is a large company or organization who specifically chooses the firm because of the quality, competitiveness and reputation of the construction work, knowing that the firm will then employ a designer to meet the brief. Such a client will issue a performance specification, outlining the functions, activities and aesthetic principles but leaving the detailed design to the contractor's designer.

Figure 12.2 sets out both the conventional contractual relationship (with independent designer) and the design and build approach as a diagram. You will see that the conventional approach ensures that the designers are independent because they are commissioned by and have a contract with the client, and have no contract with and receive no payment from the contractor. With design and build, the designer has no contract with the client but instead has a contract with and is paid by the contractor. In the examples above the contract is between the client and the contractor but there is no contract between the client and the designer. There is, however, a contract between the contractor and the designer, which may be a contract of employment if the designer works in-house. The client may liaise with the contractor's designer in the same way that they would an independent designer, but a designer tied to a contractor can never be counted upon to provide truly independent advice. There is always the risk that materials and workmanship specified or offered will be geared to what the contractor can do best or on which it can obtain the best margin of profit, rather than the best possible advice. Such compromise is not

FIGURE 12.2
CONTRACTS

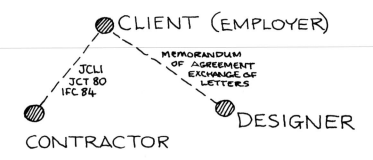

CONTRACTS

CONVENTIONAL APPROACH
INDEPENDENT DESIGNER.

CLIENT (EMPLOYER)

JCLI
JCT 80
IFC 84

MEMORANDUM
OF AGREEMENT
EXCHANGE OF
LETTERS

CONTRACTOR

DESIGNER

certain but the risk must be understood. If, for example, you approached a financial adviser tied to a bank, you might expect to hear about the performance of investments and pensions sold by that bank but not to be told the full story of how they performed compared to others. Indeed, such advisers will often show you their fund with the best rating in the league table but not show you that it is performing less well than other funds that are lower ranked. An independent financial adviser (now that they are regulated) will select the best performing firm for the best performing type of fund. It is for this reason that it is always preferable to seek truly independent professional advice.

Fees the designer charges for design and build projects would not (normally) be paid by the client as there is no contract between them. These fees would be paid by the contractor, who would then recover the cost of the design fees within the cost of the works. However, the designer will most often take instruction directly from the client in preparing the designs, although he or she would be contractually bound to check each part of the design with the contractor for any financial or technical implications it might have for plant, labour and materials required.

With small design and build companies there is always a risk that the client will take the free design advice offered and then decide not to proceed with the scheme. The designer will need to be paid for his or her time within 30 days or so of completing the work or receive wages if employed in-house. This cost then has to be covered by the contracts that do proceed. There are always clients who wish to take advantage of the free design advice but seek competitive tenders for the work. The landscape design firm must protect itself by being prepared legally to defend the copyright on designs, and many firms have terms that state that designs agreed with their designer that do not lead on to works must be paid for on a time-fee basis.

In design and build contracts the contractor will pay the designer in a variety of ways – by salary, salary and commission, percentage fee on completion of the design (often half paid on completion of works) or time-based fee to pre-agreed ceiling figure.

Subcontractors for design and build projects

Subcontractors for design and build projects may well be employed by the main contractor, and often the designer will be on a subcontract. With design and build, however, subcontractors will invariably be domestic subcontractors, chosen and employed by the main contractor. Only where a client has a specific item of specialist work, such as a sculpture, where the service provided is either unique or very rare, will the contract between the client and the contractor have provision for a nominated subcontractor. In such cases a percentage addition to the cost of the specialist works would be due to the main contractor for attendance and furthermore provision could be made for direct payment to the nominated subcontractor in the event of non-payment by the main contractor. The client (employer) would be liable for errors and omissions by the nominated subcontractor.

CHAPTER 13

Contract administration

ONCE the contractor has been chosen contract administration starts. It begins with the pre-contract meeting and ends when all contracted parties have fulfilled their obligations. The steps of the procedure are set out below:

1.   The pre-contract meeting.
2.   Variations and instructions.
3.   Valuations and certificates of payment.
4.   Progress and other certificates.
5.   Snagging.
6.   Making good defects.
7.   Certifying final completion and hand-over to the employer.

The pre-contract meeting

This vital meeting is held between the client and the contractor (often before the official signing of the contract but after the contractor has been notified that its tender has been accepted). It is the most important meeting of the scheme because it sets the scene, and any operational problems or specific points of the specification can be highlighted.

At this meeting the designer has the opportunity to point out difficult aspects of the job, areas where special care are needed (e.g. existing roots close to the surface). You can specify the works that you wish to be on site to inspect, for critical operations or those often poorly carried out. You can request material samples for approval, discuss methods of setting out and working practices and arrangements. Also it is important to sort out the location of site accommodation, access points, material and spoil heaps and how the site will be kept secure. Dates for possession will be determined, as will the completion date. The programming will be discussed and the contractor will waste no time in pointing out any deemed variations required or errors in the bills. The dates for site meetings and valuations and the method for all instructions and variations are also determined. A typical agenda for a pre-contract meeting is set out below.

Agenda for pre-contract meeting

1. Contact names/addresses/telephone numbers.
2. Dates for possession and completion.
3. Copies of contract documents.
4. Procedure for issuing instructions, variations and certificates.
5. Site access, services, signs/boards, security.
6. Hours of working and noise restrictions.
7. Site personnel, accommodation and welfare requirements.
8. Dates for regular progress meetings, site inspections and valuations.
9. Special items requiring particular attention/supervision.
10. Samples, quality control, inspections.
11. Contractor's programme.
12. Any other business.

Variations and instructions

The first variations (additions to and omissions from the scheme) required will be pointed out at the pre-contract meeting by both the designer and the contractor. The designer may have noticed errors too near the end of the tender period to have issued a circular notice to all tenderers or perhaps there are items on site that have only just come to light, such as recent damage to an existing path to be retained or a tree that has to be omitted due to a change in direction of a proposed drain.

The contractor will have checked the contract documents very carefully during the tender process and will have spotted any discrepancies or errors in the quantities. This stage is where the contractor points out any items that are under-measured but which cannot be omitted. These are called deemed variations. Clever contractors will have exploited such errors in their tender by increasing the price of under-measured items and reducing the price of any over-measured items, so that the variations that the designer must necessarily authorize will add to the contractor's profit. This is why it is vital to take off measurements for tender documents as accurately as possible.

Further additions and omissions will be required to carry out the scheme as events unfold. The contractor may, for example, find a previously unknown concrete base below the soil in an area to be planted, which will require breaking out. The contractor will not, of course, have priced for this work and a variation – adding for this work – will be required.

Any minor change in the design as a result of unforeseen occurrences that affects the quantities of the various items of work will require variation, adding or omitting the works as necessary. Such variations will be determined at regular site meetings and then written up on to landscape designer's instructions or supervising officer's instructions if in a local authority (generally known as and referred to as 'AIs' – derived from Architect's Instructions). These are set out to a standard format and a pre-printed form is supplied in pads by the

FIGURE 13.1

INSTRUCTION FORM

Landscape Architect's name and address

Works

situate at

To contractor

Under the terms of the Contract

The Landscape Institute

Instruction

for Landscape Works

Instruction no.

Date

I/We issue the following instructions. Where applicable the contract sum will be adjusted in accordance with the terms of the relevant Condition.

Instructions | £ omit | £ add

Office reference

Signed _____

Notes

Amount of contract sum £
± Approximate value of previous instructions £ _____
£
± Approximate value of this instruction £ _____
Approximate adjusted total £

Distribution: Client ☐   Contractor ☐   Quantity surveyor ☐   Clerk of works ☐   File ☐

© 1988 The Landscape Institute

respective institute (RIBA or LI). Figure 13.1 shows the pre-printed form; Figure 13.2 provides an example of completed instructions.

Sometimes it is necessary to issue an instruction to carry out work on site that does not require a variation. Such an instruction might be to move tree positions from that on the drawing because of a re-routed service run, for example. There will be no additional cost for this, assuming that the contractor has not already dug the pit. Such instructions will be also written on AIs but, having no cost implication, the initials NCE will be written next to the instruction, standing for 'no cost effect'.

Any oral instructions given on site by the designer must be confirmed in writing within two days, and from seven days of receipt of notice by the contractor that compliance with an instruction is required, the designer has the right to employ others to carry out the works, charged against sums owing to the contractor.

Valuations and certificates of payment

Valuations are carried out by the designer (although the designer is now called the contract administrator). On large projects this role might be shared with or carried out solely by a quantity surveyor. The contractor is always present and will try to ensure that as much of the work undertaken is covered by the valuation, while the designer tries to ensure that the employer does not pay for works that could be defective. The valuation is carried out by assessing the percentage of each element of the works satisfactorily completed by the contractor. This may be 100 per cent of the demolition work but only 50 per cent of the planting, 80 per cent of the brick edging and so on. The relevant percentages of the sums in the contract documents are recorded and the total amount of these sums is then written out on a certificate of payment (again these are available in pre-printed pads [see Figure 13.3]), which provides four duplicate copies (two for the employer, one for the contractor and one for the designer). The sum has a 5 per cent (or other percentage) retention deducted plus any sums previously certified and paid. (The retention sum is deducted and held as a contingency for snagging [see below] so that there is a sum of money at the end of the contract to ensure satisfactory completion of every last detail of the job.) The contractor must then send the employer a VAT invoice and the employer must then pay the contractor the amount shown on the certificate, plus VAT, within 14 days.

Progress and other certificates

The contract administrator also certifies key stages in the contract. The first of these certificates is that of practical completion, issued when the works are substantially and in all practical purposes complete (see Figure 13.4). The site does

Figure 13.2   COMPLETED INSTRUCTIONS

Issued by:	As above	**Serial no:**	02.
Employer:	xxxxxxxxxxxxxxxxx		
Address:	xxxxxxxxxxxxxxxx Hotel		
		Job ref:	JBA 92/47
Contractor:	xxx Landscape Construction		
Address:	Barley Barns, xxxxxxxxxxxxx	**Issue date:**	08/04/98
	xxxxxxxxxxxxxxxxxxxxxx		
Works:	HOTEL CAR PARK/ COURT.	**Contract dated:**	23/09/97
Situated at:	xxxxxxxxxxx Road, Cambridge.		

Under terms of above contract, I issue instructions as follows:
Office use: approx costs

		£ omit	£ add
Add for the following:			
All quantities and prices marked * are			
provisional and subject to final checking/			
agreement with the Supervising Officer.			

No.	Description	Qty	£ omit	£ add
1.	Excavation and removal of contaminated soil.	4.5 m²		£3 647.70
2.	Trenches for ducting – extra over billed items.	194 m*		£886.58*
3.	Bypass separator – extra over.	1 No.		£2 638.10
4.	Adjusting existing manholes – extra over.	2 No.		£103.50
5.	Supply/install Birco channel.	7.5 m		£947.20*
6.	Stopcock boxes.	4 No.		£474.32*
7.	Inset manhole covers – provisional item.	a) 16 No.		£602.08
		b) 1 No.		£86.93
8.	MDPE water pipe.	109 m*		£377.14*
9.	Ducting for lighting.	114 m*		£573.12*
10.	100 mm dia. ducting for CCTV.	71 m		£610.60*
11.	Supply of Blockley kerb (extra over BDC).	67 m		£1 223.42
12.	Sett edging to staff house including extra over for lifting and relaying existing conc. blocks and cutting in.	11 m		£416.59
13.	Extra over 80 mm BDC blocks in lieu of 60 mm specified (supply).	216 m²		£414.20
14.	Extra over for lean mix – BDC paved areas.	660 m²		£7 068.60
15.	Extra over for lean mix – Blockley paved areas.	590 m²		£4 926.50
16.	Extra over for lean mix – sett paved areas.	246 m²		£307.50
17.	Extra over for supply of pink setts to infill areas, over grey specified.	228.5 m²		£2 968.22
18.	Supply yellow stock bricks to client's approval. Prov. item.	3500 No.		£1 600.00
19.	Bollards – extra over bill.	5 No.		£1 227.25
20.	Lighting column – 70 w SON lamps and fittings.	8 No.		£832.00
21.	Supply of red engineering bricks for raised beds – extra over.	29.5 m		£386.75
22.	Construct brick walls for raised beds – extra over.	29.5 m		£1 443.73

Page 1

Figure 13.2 COMPLETED INSTRUCTIONS (concluded)

23.	Omit 20% supply setts a) Edging	242	£1 018.34	
	b) Channel	138 m	£569.11	
24.	Add for 1.0 m trench as bill ref. 2.3B on pages 4/5.	12 m		£179.88
25.	Omit 0.5 m trench as bill ref. 2.3A on pages 4/5.	3 m	£35.79	
26.	Omit 1.5 m trench as bill ref. 2.3A on pages 4/5.	31 m*	£849.40	
27.	Omit excavation for gulleys.	3 No.	£57.90	
28.	Omit excavation for inspection chambers.	1 No.	£52.33	
29.	Omit gulleys, bill ref. 3.1.	3 No.	£678.36	
30.	Omit inspection chamber, ref. 3.2.	1 No.	£389.87	
31.	Add for 150 mm dia. pipe bill ref. 3.4A.	88 m		£2 946.24
32.	Omit 225 mm dia. pipe as ref. 3.4B.	110 m	£4 235.00	
33.	Add for capping existing gulleys as bill ref. 3.6 pages 4/8.	2 No.		£184.00
34.	Add for works to existing manholes as bill ref. 3.6 pages 4/8.	2 No.		£103.50
35.	Add for draw pit for lighting.	1 No.		£120.00*
36.	Omit supply and lay of granite sett – bill ref. 4.3.	67 m*	£2 537.39*	
37.	Add for supply and lay of granite sett and BDC edging – all as bill ref. 4.3.	78 m*		£2447.64*
	COLLECTION OF ADDS AND OMITS		£10 423.49	£39 743.29

Signed _____

Amount of contract sum	£193,341.68
+/– Approx value of previous instructions	–£48, 924.00
	£144,417.68
+/– Approx value of this instruction	+£29,319.80
Approximate adjusted total	£173,737.48

Distribution: CLIENT, CONTRACTOR, FILE.

Page 2

not have to be totally complete. There can be small, insignificant matters outstanding or at fault, which are called 'snags'. This certificate entitles the contractor to receive half the retention money. When the retention is 5 per cent, then clearly at practical completion a payment certificate is required (which must be issued within 14 days of issuing the practical completion certificate) authorizing 97.5 per cent of the contract sum, less any sums set aside for maintenance.

Practical completion heralds the start of the 'defects liability period' (the period of time that the contractor is liable for defects) and it also heralds the commencement of the maintenance period. The contractor is no longer responsible for site security and theft or vandalism, or for insuring the site.

FIGURE 13.3
VALUATION FORM

Landscape Valuation/Financial Statement   No.

Landscape
Architects

Valuation

Employer:
address:
Contractor:
address:

Contract   Office reference

AUTHORISED EXPENDITURE   **ESTIMATED FINAL COST**

To amount of Contract   £ _____   To amount of Contract   £ _ _ _ _ _ _ _

Additional authorised expenditure   £ _ _ _ _ _ _ _   *Less* contingencies   £ _ _ _ _ _ _ _

                                                                              £ _ _ _ _ _ _ _

                                        Estimated value of Variations to date   +£ _ _ _ _ _ _ _

                                                                              £ _ _ _ _ _ _ _

                                        Estimated value of Variations yet to   +£ _ _ _ _ _ _ _
                                        be issued

                                                                              £ _ _ _ _ _ _ _

Estimated increased costs (if allowable) £ _ _ _ _ _ _ _   Estimated increased costs (if allowable) +£ _ _ _ _ _ _ _

TOTAL AUTHORISED EXPENDITURE £ _____   **ESTIMATED TOTAL COST**   £ _____

Signed _____   Landscape Architect. Date _____

Distribution: Client ☐   Quantity surveyor ☐   File ☐   ☐   ☐

Landscape Interim/Final Certificate   No.

Landscape
Architect:
address:

Landscape
Architects
Interim/Final

Certificate

Employer:
address:

Certificate No.   _____

Date of Valuation   _____

Date of Certificate   _____

Contractor:
address:

Valuation of work
executed £ _ _ _ _ _ _ _

Value of materials
on site £ _ _ _ _ _ _ _

                                                                              £ _ _ _ _ _ _ _

Contract:

Reference: *Less* retention £ _ _ _ _ _ _ _

I/We hereby certify that under the terms of the Contract   Total to date £ _ _ _ _ _ _ _

                                                        Less amount previously
dated: _____   certified £ _ _ _ _ _ _ _

the sum of (words) _____   **TOTAL**   £ _____

_____   (Exclusive of any Value Added Tax)

is due from the Employer to the Contractor

Signed _____   Landscape Architect: Date _____

Distribution: Client ☐   Contractor ☐   Quantity surveyor ☐   File ☐   ☐

Landscape Architect
address

Employer
address

Works
situated at

Contractor
address

Contract dated

Certificate of

Practical Completion

of Landscape Works

Job reference

Serial No.

Issue date

Under the terms of the above mentioned Contract,

I/We certify that Practical Completion was achieved and the works taken into possesion on

_____ 19 _____

the Defects Liability period
for faults other than plant failures will expire on

_____ 19 _____

for shrubs, ordinary nursery stock trees and other plants will expire on

_____ 19 _____

for semi mature and advanced nursery stock trees will expire on

_____ 19 _____

The Employer should note that as from date of issue of this Certificate of Practical Completion of the Works the Employer becomes solely responsible for insurance of the Works.

To be signed by or for the issuer named above.

Signed _____

Date _____

Distribution: Client ☐   Contractor ☐   Quantity surveyor ☐   Clerk of works ☐   File ☐

© 1987 The Landscape Institute

FIGURE 13.4
PRACTICAL COMPLETION
FORM

If the contractor has not completed the works by the date when completion was due, then the designer must issue a certificate of non-completion (see Figure 13.5). In this case (when an extension of the contract period has not been agreed) liquidated and ascertained damages become payable each week, calculated to a formula proven to stand up in court, which compensates the employer for notional loss of capital. The following items must be taken into account when considering the total sum payable per week.

1. Notional loss of capital.
2. Inconvenience and any actual consequential loss.
3. Additional professional fees.

The established formula for calculating liquidated and ascertained damages for landscape contracts is set out below:

$$\frac{\text{Contract sum} \times \text{bank rate} + 2\%}{52 \text{ weeks}} = £ .... \text{ per week}$$

e.g. Estimated contract value £75 000

Bank rate + 2% = 9%

$$\text{Interest} = \frac{£75\,000 \times 9\%}{52} = £129.80$$

To this sum should be added any consequential loss and any likely additional professional fees. The consequential loss for a drive-in fast food outlet might be as much as £4000 a day if the delay in finishing the works prevented the outlet opening for business. Apart from such situations, such loss is not common in landscape schemes.

Snagging

The contract administrator produces a list of all defects and outstanding items when issuing a certificate of practical completion – called a snagging list. The designer agrees with the contractor a time when all the snags are to be satisfactorily resolved. The snagging list will also include a dead count of plants, though these may be replaced at the end of the growing season following planting. Contractors sometimes refer to this stage not as snagging but as 'nagging'. Some contractors are better than others at clearing up such snags and therefore there are always some who require much more nagging from the contract administrator.

Making good defects

When the defects are made good and at the end of the defects liability period,

FIGURE 13.5 NON-COMPLETION FORM

Landscape Architect
address

Employer
address

Works
situated at

Contractor
address

Contract dated

Certificate of

Non-completion

of Landscape Work

Job reference

Serial no.

Issue date

Under the terms of the above mentioned Contract,

I/We certify that the Contractor has failed to complete the Works by the
Date for Completion or within any extended time fixed under the contract
provisions.

To be signed by or for
the issuer named
above.

Signed _____

Date _____

Distribution: Client ☐   Contractor ☐   Quantity surveyor ☐   Clerk of works ☐   File

© 1987 The Landscape Ins

the designer should issue a certificate of making good defects. The balance of the retention money can now be released. If the defects are not made good by this time there is another certificate for 'Non-Making Good Defects', and retention money will be retained until such time as the defects are made good.

It is essential with small landscape schemes to ensure that the retention is sufficient to make it worthwhile for the contractor to return to site to clear up snags satisfactorily. Some will not, writing off small retentions. If this happens you can advise the contractor that you will remove them from the approved list of contractors unless the work is carried out or else, having first issued a written instruction to carry out the works, and seven days having elapsed, the contract administrator can employ another firm to carry out the work and recover the cost of this work from the contractor as a debt, or deduct it from sums owing.

Certifying final completion and hand-over to the employer

Within three months (or other stated period) of practical completion, the contractor must supply to the contract administrator (designer or quantity surveyor) all documentation required to prepare a final valuation. When this information has been obtained, the defects liability period (including for plant replacement) has ended, any concurrent maintenance period has ended and all defects have been made good, the contract administrator must, within 28 days, issue the final certificate – releasing the balance of the money due to the contractor. The contract is then completed and sometimes celebrated with a formal hand-over meeting. The employer then takes over the responsibility of maintenance.

Site inspections and administration

It is not easy to administer a contract and tread what is a tightrope between being viewed as picky, petty and ignoring advice, and being a soft touch who will be swayed into accepting cheaper alternative materials and practices. Obviously the proper path lies somewhere in the middle. The inexperienced contract administrator will tend to lean towards one of these extremes, but with increasing experience he or she should soon learn to discriminate good advice from corner-cutting, and know when it is necessary to get tough.

However, if you wish to realize the dream that you started with back at the thumbnail sketch stage and assuming that all the specification notes contained on your plans and schedules are put there for a practical and workable purpose, then it is essential that you ensure that they are carried out to the letter by the contractor – by careful site inspection and administration. If the design is good and the specification is strictly followed and policed, a very attractive landscape will result. Strict adherence to the specification by close inspection guarantees that each contractor is competing equally and fairly; the lowest tender being the most competitive for the same end result.

Clearly, all contractors should be made to carry out the specification to the letter or else they would be able to reduce their tender bids, knowing that a lesser, cheaper product would be accepted. Such conduct not only cheats the conscientious contractor but also cheats your client of a good result and, moreover, of good value for money.

Without rigorous inspection value for money is lost for two reasons. First, the contractor will always reduce its tender price by a lower amount than the value of the works it is able to omit or water down. Therefore the rate such contractors get for the value of work they actually do may be higher than for a conscientious contractor who priced the work exactly to the specification. Second, the saving in the tender price made by the client in allowing a poor job of work is a small percentage of the overall landscape cost and yet will make a dramatic difference to the end result – the difference between the success and the failure of the scheme.

Some contractors may find all manner of excuses for failing to perform to the specification but few (if any) of them stand up to serious scrutiny. Below are listed examples of such excuses and details of how notes in the specification can avoid them being used:

1. *Plants of the species and size are not available, requiring substitution with smaller-sized plants.* Counteract this common excuse by a clause such as 'Non-availability of species or size of plant to be notified to the designer before tender submissions so truth of this can be determined and a substitution can be made if necessary and all tenderers advised accordingly by circular letter.'
2. *Plants were stolen or damaged by the client, the client's customers or pets.* Counteract this excuse by making the necessary options in the JCLI contract and/or adding in the notes: 'Any plant deaths to be at the contractor's own risk during the entire term of the contract for the landscape works. The contractor shall allow in its tender price for any such replacements that occur, whether due to damage in planting, incorrect planting depth, insufficient water, insufficient ground preparation, theft, vandalism, urine from household pets or any other similar cause, and to include damage caused directly by the employer's customers, staff or subcontractors.'
3. *Climbers did not require supports or residents refused to accept climbers and/or training supports on their walls.* This excuse can be counteracted as follows: 'Climbers shall be planted as specified to include training wires (necessary to anchor stems to wall so that wind movement does not prevent plant clinging to wall) and trellis panels (necessary with spacing blocks to allow twining climbers to twine around panel). Where individual residents expressly object to climbing plants, the contractor shall make due explanation of the planning requirements, pointing out that the supports proposed will cause no damage to their wall. If after such explanation the customer objects to the climber, then the plot number shall be noted and the devel-

oper credited with the supply and fixing cost of the climber on a plot by plot basis.'

4. *We cannot water the plants on planting because it is illegal to use the hydrants.* Counteract this excuse by: 'The contractor shall water in all plants as specified, and shall be responsible for contacting the site agent (if there is a main contractor) and agreeing a suitable water supply point or else contacting the local water authority for a suitable licence for the duration of the contract.'

There are always occasions when an excuse is genuine, or when advice tendered is sound and helpful, and the better contractors can provide helpful practical advice on how best to overcome problems that inevitably crop up on site. After all, the client is using a specialist landscape contractor because he or she possesses specialist skills and expertise. The real test of a good designer is the ability to determine when to take advice and when to respectfully reject it, when to accept a substitution and when to stick to your guns and defend your specification. Such a designer will consistently achieve good results.

Maintenance factors

While there seems to be ever more pressure to design low maintenance schemes, using plants that need minimal pruning and mulching specifications that greatly reduce weed and watering requirements, the fact remains that the maintenance period can be the making or breaking of many landscape designs. The first year is a crucial stage of establishment and initial growth, and some plants will establish well and fast, while others struggle. This may, of course, be a result of poor plant choice but often it is simply a quirk of the site. A normally vigorous shrub may for no obvious reason struggle or be slow to gain momentum. During such time normally smaller shrubs in front might temporarily outgrow taller-growing varieties behind. Such occurrences can be due to local waterlogging, pests, poor stock, local dry spots, microclimatic factors, accidental or deliberate damage or whatever.

The defects liability period is essential to ensure that any inadequacies of the contractor's workmanship, such as poor ground preparation, can be recognized and put right. However, unless the contractor is responsible for the initial one (or two) years' maintenance, any defective plant stock or plants damaged in planting could be blamed by the contractor on the client's poor aftercare and it will be hard to provide evidence to the contrary. It is therefore also essential that the contractor be responsible for maintenance for at least as long as the defects liability period, and where shrubs and trees are concerned both these periods should be for a minimum of 12 months.

A contractor's maintenance period will serve to mitigate the effects of variable site conditions and plant establishment. In order to cover all works necessary to put right problems on site, provision must be made for these works in clear and precise clauses, not only describing the nature of the work but also quantifying, as far as possible, the number of visits required.

Cost

By quantifying the number of maintenance visits and the nature of the works carried out at each visit, a contractor takes less risk in pricing for these items in advance at the tender stage. The more open-ended the maintenance clauses, the more risk a contractor takes or the higher the price tendered for maintenance. This may result in an otherwise good and competitive contractor losing the contract.

It is easier to foresee the maintenance requirement for a 12-month period than for 24 months, and there is correspondingly less risk for a contractor for the shorter period. Where 24 months is necessary because of the larger nursery stock trees planted, it may be fairer (and also may ensure lower tender prices) if a provisional sum is set aside for expenditure when the maintenance works are required, instructed and satisfactorily completed. It is then the designer's responsibility, when preparing the tender documentation, to gauge how large this sum should be to cover the likely maintenance. This provisional sum method at least ensures that only what is actually necessary is expended.

The risk in pricing for such items as watering, weeding, pruning, mowing, herbiciding, pest control, tidying and replacement is not too great for a contractor if limited to one year (or at worst two years), so long as the visits are quantified, but longer term the responsibility of ongoing maintenance should not be placed upon the contractor implementing the scheme at the time of tender. Such a contractor would not only have to guess the level of input required but also have to allow for inflation and likely increases in overheads. This might cost the contractor dear or, more likely (in order to cover this eventuality), it might result in a high tender price. For these reasons, any maintenance requirements after the initial establishment period must be separately tendered and such works need separate documentation, set out to provide a thorough management plan for all aspects of the landscape scheme.

The management plan

A management plan can also be prepared for a client or the client's gardener. The specification of ongoing maintenance works will have to be far more comprehensive, however, if it is to be tendered to maintenance contractors, in order to tie down every detail of the materials and workmanship required. This detail is less important if a client (or the client's gardener) takes over responsibility for maintenance. After all, it is the competition between contractors that provides the incentive to cut corners to win the work and this factor demands pedantic clarification of every last detail.

The management plan will comprise both a drawn location plan and a full descriptive document concerning the works (if a tender document, this will include a preliminaries section, specification, schedule of provisional sums and a measured works section). Set out below is an example of the most common maintenance clauses, though the level of specification represents merely a sample and is not intended as a comprehensive guide to all aspects of maintenance.

Indeed, the following clauses might be most often used for the first 12 months' maintenance rather than for a full plan of ongoing maintenance works, without limit.

Sample maintenance clauses – first 12 months

Generally maintenance shall commence from the completion of works on site and shall cease 12 months following the completion date. The contractor shall be responsible for the timing of visits to ensure that the site has a well-maintained appearance. The contractor's maintenance rate shall include for all the following items:

Weeding

Remove all weed growth by hand as necessary to ensure weed-free and tidy planting beds. Take great care not to disturb sheet or bark mulch. All weeds shall be removed from the site. Two visits are required per growing season where sheet mulch is specified, and six visits per growing season are required where no sheet mulch is specified. Visits should occur approximately monthly during the growing season, subject to weather conditions.

Spot herbiciding

Persistent perennial weeds shall be controlled using a suitable herbicide to the manufacturer's instructions, as required and at intervals to ensure no regeneration of weed. Extreme care must be taken to avoid damage to surrounding plants and grass, and to avoid spray drift. Any herbicide treatment shall comply with the Control of Pesticide Regulations 1986 and all relevant COSHH Regulations, and the type of herbicide shall first be agreed with the landscape architect. Spraying shall be carried out by a qualified and skilled operative and shall not be undertaken in windy conditions. All damage resulting from incorrect usage, spillage and spray drift to be rectified at the contractor's expense.

One application is required per growing season where sheet mulch is specified, and four visits per growing season are required where no sheet mulch is specified. Visits should occur approximately every six weeks, subject to weather conditions.

Watering

Shrubs and trees

Where sheet mulch is specified, allow only one watering visit during any spell of continuous hot weather lasting more than 14 days at the rates given below. Where sheet mulch is not specified, water all shrubs to field capacity (minimum 25 litres per tree position and 20 litres per m^2) on ten separate and agreed occasions during the maintenance period.

Turf

Water turf to field capacity (minimum 20 litres per m^2) on ten separate and agreed occasions during the maintenance period.

Watering generally

Watering shall be carried out during dry periods (less than 30 mm rainfall at end of any four-week period) and the contractor shall be responsible for determining such dry periods and for contacting the landscape architect and agreeing the timing of each watering visit. The contractor shall replace at its own expense any planting or turf that fails due to lack of water.

During periods of drought, when restrictions are placed on the use of water, the contractor shall be responsible for notifying the landscape architect of sources of second-class water and the costs of obtaining such water.

Shrubs, trees, tidying beds and mulch levels

Remove all litter and debris at each visit, leaving the site clean and tidy. Firm in and straighten any plants loosened and prune out dead, leggy and broken branches, without damage to natural habit of plant. In the case of trees such pruning shall be carried out by a suitably skilled and qualified arboriculturalist. Tree stakes and ties shall be checked, adjusted and replaced as necessary. Prune hedges back to even hedge line to encourage thickening twice within the first growing season after planting. All trees and shrubs shall be fertilized using an approved liquid feed (N10:P15:P10) at a rate of 60 g/m^2 during early May and late September. Top up mulch levels to a minimum of 50 mm and ensure that all loose mulch is raked back on to beds from any hard or grass surfaces. Peg any loose flaps of sheet mulch.

Control of insects and diseases

Control insects, fungus and other diseases by spraying with an approved insecticide or fungicide. Pest problems and methods of control shall be agreed in advance with the inspecting landscape architect. Any pesticide treatment shall comply with the Control of Pesticide Regulations 1986 and all relevant COSHH Regulations, and the type of herbicide shall first be agreed with the landscape architect. Spraying shall be carried out by a qualified and skilled operative and shall not be undertaken in windy conditions.

Mowing

Mow seeded areas [as set out in the seeding clauses above] and continue to mow as follows:

Sports area

Mow to a height of 20 mm, at 14-day intervals throughout the growing season, and allow 14 cuts in total. Remove grass cuttings from site after each visit. Trim edges and mowing margins. Roll grass in April, June and August. Apply approved turf fertilizer, selective weed killer and moss retardant in May and September.

General amenity areas

Mow to height of 25 mm, at 14-day intervals throughout the growing season, and allow 14 cuts in total. Remove grass cuttings from site from May to August. Strim edges and mowing margins. Roll grass in April, June and August. Apply approved turf fertilizer, selective weed killer and moss retardant in May and September ensuring compliance with the Control of Pesticide Regulations 1986 and all relevant COSHH Regulations. Spraying to be carried out by a skilled, qualified and certified operative.

Wild flower area – summer flowering
Strim to height of 75 mm, once in late August, once again in mid-October and in mild winters once again in mid-March. Rake off all cut material five days after each cutting and remove from site.

Wild flower area – spring flowering
Strim to height of 50 mm, once in late June, once again in mid-August and once again in early October. Rake off all cut material five days after each cutting and remove from site.

Replacement

Replace at once any plants that before the end of the maintenance period fail to show growth or develop full foliage during the growing season after planting (including plants damaged during maintenance operations) but do not include any plant that after practical completion is destroyed by vandalism, theft or similar cause through no fault of the contractor. The opinion of the inspecting landscape architect shall be final in judging damaged and unhealthy plants to be replaced. Shrubs and trees so replaced shall be the same or similar to those specified, previously supplied and approved. All such replacement planting to be at the contractor's expense, including any works necessary to enable planting to be properly carried out, i.e. removal and disposal of dead material. A full snagging list including all defective preparation, construction and plant stock shall be prepared by the contractor in the presence of the inspecting landscape architect at the end of the growing season following planting (around 30 September).

Notification

The contractor shall notify the contract administrator of all visits, and shall provide written attendance sheets detailing time, date, works carried out, persons on site notified. No certificate will be issued for maintenance visits without prior receipt of an attendance sheet.

CHAPTER 14

Graphic presentation, draughting equipment and plan printing

THE subject of graphics, equipment and draughtsmanship is complex and therefore for detailed study the reader is referred to the specialist works. The following passages offer an introduction to the subject.

Graphic equipment

The main items that are used daily in design offices are listed below. Any references to brand names are merely illustrative of the type and grade of material required. This list is not meant as an ABC of every tool, graphic aid and piece of equipment made, but as a pragmatic guide to equipping an ample, suitable and cost effective work station for most types of scheme and situation.

Pens

0.18, 0.25, 0.4, 0.6, 0.8 and 1.2 Rotring™ or Faber Castell™ drawing pens. The absolute minimum number of pens for hard up students is the 0.18, 0.4 and the 0.8, though the 0.25 is also very useful. These pens are used for drawing lines on tracing paper to show the various elements of the drawing. Read the instructions carefully. Clean nibs regularly to avoid blobbing. Do not leave unused, filled with ink, in a warm room. Do not overfill.

Ink and pen cleaner

Rotring or Faber Castell to match pens. Don't mix the types as Rotring ink in a Faber Castell pen will clog it. It is best to fill the cartridge three-quarters full. Include a pen bucket for when your pens clog up – soak them well overnight or if really bad take the pen apart and soak the pieces. Then rinse well.

Clutch pencil

0.5 with leads – suggest H. A clutch pencil is a plastic holder in the shape of a pen which has a clasp at the head, which holds on to the lead. As the lead wears down you release the clasp and allow the lead to drop further. This avoids constantly having to sharpen your pencil. These pencils are for sketching out. Place 5 cm long leads in the end and these feed down.

Erasers

A pencil eraser and an ink eraser: Rotring is a good make. Beware that rubbing out is not accurate and can take out parts of the drawing you wished to retain. Clean rubbers by rubbing them on a piece of carpet or wood. An old offcut is ideal.

Razor blades

Several are needed – sharp ones! Take great care. These allow accurate erasing of ink on tracing paper. Use the edge to scrape surface. After the lines are removed, rub with an eraser and rub a finger nail over affected area to smooth down any rough scales. Unless this process is carried out the pen will draw a line much thicker than previous lines drawn with it, because the rough surface will cause the ink to spread.

Graph paper

A1 size piece of metric graph paper with 1 mm squares. This is a very useful piece of equipment. Adhere the graph paper to the drawing board when you draw on tracing paper (most commonly used for drawing work), then the lines of the graph paper can be used as a guide for all freehand work, whether drawing lines (especially horizontal and vertical lines) or lettering. The lines will help you to ensure that all architectural lettering is level, aligned and of the correct size. If you are 'taking off' quantities, the squares can be used – $0.5 \text{ cm}^2 = 1 \text{ m}^2$ at $1:200$. $1 \text{ cm}^2 = 1 \text{ m}^2$ at $1:100$.

Masking tape

Use pieces about 3 cm long for holding down the plan edges. If you get trouble with the paper slipping, use a larger piece. If a tracing paper negative is left overnight it may be necessary to re-fix it, as these can absorb water vapour and expand.

Scale rule

With scales $1:100$, $1:200$, $1:500$, $1:1250$, $1:2500$. Also useful (but not essential, as you can use the scales above and divide by 10) are scales $1:50$, $1:10$,

1:5, 1:20. These scales are used for measuring lines to accurate scales. If scaling up, measure each line at (say) 1:200 and then measure and draw the line on the new plan at (say) 1:100. Clean scale rulers by rubbing edge along a dry cloth.

Set square

Adjustable are best – the bigger the better. For drawing lines at measurable angles. Use on parallel motion to get parallel diagonal lines. Clean by rubbing edge on a cloth.

Drawing board

With parallel motion – preferably A0 for comfort and versatility. Use parallel motion for horizontal lines, and stand the set square on the parallel motion for vertical lines. Fix a piece of metric graph paper to the face and replace this regularly. Clean the board daily before use. However, like with many draughting professions, the more experience you have, the less equipment you need. I have seen the most eminent professionals using a piece of angled MDF and a rolling ruler.

Circle template

A wide range of circles is necessary. You may well need several sizes. For drawing neat circles of many sizes there is no quicker method. Keep ink pen nibs away from the absolute edge of the circles or the ink will run under the template. Reduce the risk of this by using little pads of masking tape to raise the surface.

Pair of compasses or springbow, with ink attachment

For drawing much bigger circles. Ensure nib is same height as point. Draw smoothly and if pencil is used, replace blunt leads quickly.

Selection of highlighter pens

Used for 'taking off' plans. 'Take off' plans by highlighting areas, plant names and items to ensure accuracy and thoroughness.

Colour equipment

When producing design drawings for the client, sometimes drawings will require colouring. Colour can be applied to drawings using the following media. Coloured marker pens (spirit based and water based), which are used for bold blocking in of solid colours for diagrammatic drawings. Pencil crayons, which can be used for more subtle colouring of plans and perspective renditions and allow the mixing of colours to provide subtle changes of shade and tone.

Watercolour paints, which can be used for both colouring of plans and perspective renditions, are the most translucent and can (in skilled hands) produce the most striking and evocative results.

Tracing paper

A1 and A0 sheets, heavy weight 120 g. For drawing on to get dyeline prints etc. Never, never get wet, not even a drop as the paper will be destroyed. Keep rolled or else flat in a plan chest. (If rolled ensure a reasonably large diameter to reduce coil memory.) Coil memory is always a problem with rolled plans. It is the phenomenon that causes a plan to roll itself up again, no matter how hard you try to force the roll open, and paperweights never seem to be heavy enough. The tighter you roll your plans the harder they are to open out again. Mend tears quickly with magic tape before they run.

Job files and general stationery

These items are needed for running a job in an orderly fashion. Store all correspondence in date order and plans in a separate plan wallet. Tab important letters and certificates, instructions, briefing notes and fee quotations. Record and date all verbal instructions in writing, whether from the client to the designer or from the designer to the contractor.

Personal computer (PC) or Apple Macintosh™

With good up-to-date software and a good quality printer.

Magic™ tape

This is a clear tape, which is used for mending tears in tracing paper plans.

Plan printing

Negatives

The term 'negative' is given to anything from which you can take a print. We are most familiar with the term when discussing photography, but the word equally applies to the ink on tracing paper drawings prepared in a drawing office. Such negatives are used to produce dyeline prints.

Dyeline prints

Dyeline prints are produced by exposing light sensitive paper (yellow in its unexposed state) and then running the paper through a developing fluid. The light sensitive paper is placed behind the negative and the two sheets are passed through the printing machine, which shines a light through the negative. The black ink lines mask parts of the copy paper and when developed these turn from yellow to black. Other colours are available and sepia or blue prints can be produced. These paper prints will eventually fade but will last many years if not exposed to the light.

Photocopies

The increasingly popular alternative to dyeline printing is to obtain photocopies – available up to A0 size. These produce no background colour – a feature of the dyeline print – and are very clear, if a bit stark. They are still a little more expensive than dyeline prints, though many people prefer the sharpness. They do last much longer too. The additional advantage of the photocopy is that if you loose your negative, you can photocopy back on to tracing paper from a copy. Indeed, badly damaged negatives can be photocopied on to tracing paper to produce a new clean copy. The only disadvantage of the photocopy is the degree of distortion. The copied image will be slightly bigger than the original, usually in just one direction, making it impossible to correct by the photocopy reduction process.

Copy negatives

A copy negative is when a further negative is obtained from the first, sometimes referred to as 'negative to negative' printing. There are many instances when the diligent use of copy negatives will save the designer hours of drawing time. A single base plan negative may be useful to produce many working drawings, such as the hard works layout plan, the planting plan, the setting out plan and the demolition plan. By obtaining three copy negatives, the designer will save draughting the base design and all existing features to be retained three times over. When fees are tight and margins of profit low, the designer needs to make use of every trick available.

There are two ways a copy negative can be produced: first, by light sensitive film – using the dyeline process – and, second, by photocopying. The tracing paper must be of a type that is suitable for photocopies, as with untreated tracing papers the carbon easily rubs off. Again, there is more distortion with photocopying than with the dyeline method, though the sharpness of the copy is usually better.

Copy negatives are also useful to obtain a new negative from an old damaged one. The dyeline copy negative process does not allow copying back from a print on to negative film unlike the photocopying process. The advent of the A0 photocopier can save your liver and bacon (as well as your bread and butter) by allowing the copying from a print back on to tracing paper. This allows the new negative to be added to and changed to suit the changing demands of the client or site conditions.

Saving time

By using the full range of copying and printing methods outlined above, the designer can save much time, diversify plans from a single base drawing, restore damaged and lost negatives and obtain sufficient copies to supply all parties involved with the scheme, in a foldable, postable and manageable format. When preparing plant schedules and notes on drawings it can be neater, clearer and more time efficient to type up such items on a computer and print out the text. The only problem with this is that printer ink is generally unstable and when it is necessary to adhere text to a negative for dyeline printing the text should be photocopied on to adhesive-backed clear pvc (copier stable grade) and then stuck in place (cutting and placing where required). If a drawing is going to be photocopied (as opposed to dyeline printed), then text can be stuck directly on to a drawing from the printer, as long as the ink has been determined as being completely stable. If not, again resort to the photocopier but the use of adhesive pvc is not required as white paper can be cut out and stuck on with clear tape. Where a CAD drawing is plotted on to paper, the ink is commonly found to be unstable and will smudge. Further copies must therefore be produced by photocopying or (where a negative is printed) by dyeline printing. Dyeline printing is increasingly becoming outdated because of the higher quality resolution of photocopying and its improving cost-effectiveness.

CHAPTER 15

Drawing principles

CERTAIN basic principles form the foundation of the landscape designer's approach to the presentation of design concepts through the medium of drawings. These principles need to be well understood if one is to function successfully as a professional.

Graphics and draughtsmanship for plans

The following guiding principles will be of invaluable assistance when deciding which pen to use to draw a line on a plan. The correct choice of pen weight will make all the difference between a confusing series of lines and a coherent plan with clear portrayal of depth (three-dimensional effect), showing each element clearly in relation to the others.

The bird's-eye view principle

The relative width of drawn lines determines the appearance of height of the drawn object. Thick lines make an object appear taller – in other words, using the seagull theory, thick lines appear closer to an imaginary bird flying over the site than thin lines (see Figure 15.1). Therefore if the designer wishes to draw a tall object like a building, the thickest lines will be chosen (1.0 or similar), while if the item is very low in height (and far away from the bird), such as the junction between a path and a lawn, the thinnest line will be chosen, such as a 0.18. A low wall or hedge would therefore need an intermediate line weight, such as a 0.4, and a 2 m high timber fence a 0.7 perhaps.

The narrower the distance between two lines (describing the width of an object), the taller the object appears to be. This is an optical effect, which can complicate the bird's-eye view principle just a little. For example, a free-standing brick wall might be as tall as a garage but if the same pen size were used for both items, the wall would appear taller and more dominant. To compensate for this, the wall (conventionally depicted as two parallel lines to show the thickness) would require a smaller pen size (perhaps a 0.6) than would be needed for the larger and broader garage (0.8). This rule is simple and foolproof but there are a few exceptions just to prove the rule: tree canopies, for example, on working drawings. Because proposed tree canopies are purely spec-

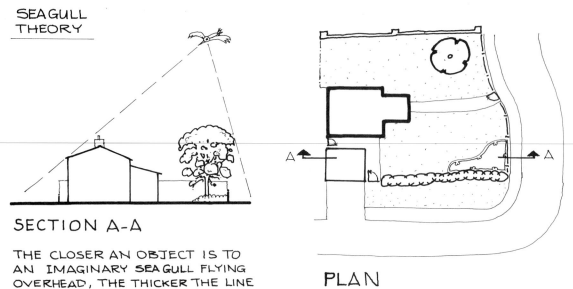

SEAGULL
THEORY

SECTION A-A

THE CLOSER AN OBJECT IS TO
AN IMAGINARY SEAGULL FLYING
OVERHEAD, THE THICKER THE LINE
HAS TO BE WHEN DRAWING THE
OBJECT IN PLAN. TREES ARE THE
EXCEPTION TO PROVE THE RULE.

PLAN

HOUSE = 1.2 PEN   EXISTING TREE = 0.7
GARAGE = 0.6   HEDGE AND FRONT FENCE
ARE 0.4 ;   BRICK WALL = 0.4. SHRUBS
ARE 0.25 AND PATHS / EDGE OF BED
AND GRASS TEXTURE = 0.18 PEN. REAR
FENCE = 0.25.

FIGURE 15.1 SEAGULL
THEORY

ulative or indicative, the canopy line is drawn with a thin pen, 0.2. Even large existing trees are generally given thinner lines than might be suggested by their height (0.4). This is because they are normally of secondary importance to the scheme layout and in any event they are always changing.

The change of level principle

Another situation that is not obviously explained by the seagull theory is, for example, the line to draw for the junction between a paved area and a pond or water feature. On such occasions the change of level principle can be applied (see Figure 15.2). It is necessary to grasp this principle in tandem with the seagull theory – in order to make the correct pen choice.

The change of level principle is a simple code to solve any question not fully answered by the bird's-eye view principle. It assumes that any change of level will demand a thicker pen than, say, the junction between two surfaces on a flat plane. For example, a kerb would be drawn with a thicker line than a flush edging. If you do not possess a full range of pen sizes, you can use pencil in conjunction with the pens you have available to give emphasis to different heights of objects or to abrupt changes of level. The very black pen lines will

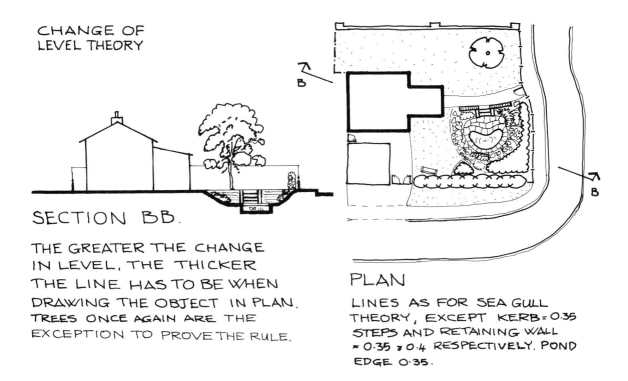

CHANGE OF
LEVEL THEORY

SECTION BB.

THE GREATER THE CHANGE
IN LEVEL, THE THICKER
THE LINE HAS TO BE WHEN
DRAWING THE OBJECT IN PLAN.
TREES ONCE AGAIN ARE THE
EXCEPTION TO PROVE THE RULE.

PLAN

LINES AS FOR SEA GULL
THEORY, EXCEPT KERB = 0.35
STEPS AND RETAINING WALL
= 0.35 & 0.4 RESPECTIVELY. POND
EDGE 0.35.

show up taller objects or changes of level against the paler pencil lines. Hard H pencils are necessary to ensure even and sharp lines and to prevent smudging. The line at the edge of the pond might be drawn in a 0.4 pen, while the line to show the junction between the pond edging and the grass (nearly flush) might be drawn in a 0.18 pen or pencil line, as would the lines for the joints between the edging units.

FIGURE 15.2 CHANGE OF LEVEL THEORY

Visual dynamics of mass and void

If you draw a vertical line on a page, say, 100 mm long, then another equal line parallel to it 75 mm away, another 150 mm away from the last and then a fourth line another 75 mm away, you may witness a curious visual effect. It often helps to draw a horizontal line along the top edge of the vertical lines and this new line can extend beyond them 100 mm either end. Another such line can be drawn along the bottom edge. Look at the lines and think of them as a plan of the ground. Now try to guess what you are looking at (see Figure 15.3). If you were told that the lines represented masses such as (buildings) and voids (such as courtyards between them), then which do you think are which? It is a curious fact that most people will state that each gap between the lines closest together represents the masses (or buildings) and the gaps between the lines furthest apart represent the voids (or courtyards).

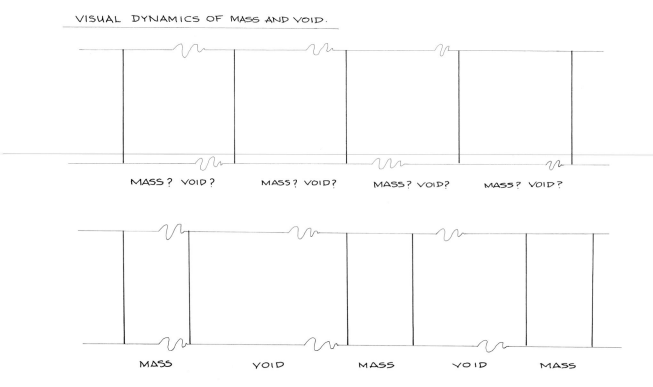

THE CLOSER TWO LINES ARE DRAWN TOGETHER (RELATIVE TO OTHER LINES) THE MORE THEY WILL APPEAR TO ADVANCE TOWARDS THE VIEWER - THE SPACE BETWEEN THEM WILL BE PERCEIVED TO BE SOLID, A MASS - OPPOSED TO THE VOID OR SPACE BEYOND. THIS IS WHY TWO PARALLEL LINES USED TO DEPICT A WALL CAN BE AS THIN AS 0.35 (AND STILL LOOK APPROPRIATE) WHEN STRICT ADHERENCE TO "SEAGULL" AND "CHANGE OF LEVEL" THEORIES MIGHT SUGGEST A 0.6 OR 0.8 PEN (WHICH IF USED WOULD APPEAR FAR TOO HEAVY).

FIGURE 15.3 VISUAL DYNAMICS OF MASS AND VOID

If a limited range of pens is available, then some variation in line can be imparted by altering the speed a line is drawn. This takes some skill and practice, as beginners will generally tend to start and finish each line they draw more slowly than the middle of the line, causing an undesirable variation in line width along its length, thicker at both the beginning and the end, but thin in the middle.

Before you can manipulate line width by varying line speed to any useful effect, you must first learn to control line speed generally. The more even the line speed for each line, the more even will be the line drawn. Once you have learned to draw each line evenly, varying the speed the line is drawn will provide a much greater variation in line weight, certainly more than might be apparent from your pen box. The faster the line is drawn, the thinner the line will be, within a range governed by the pen size. This trick cannot be so well performed with pens larger than a 0.5 and anyway is most useful for imparting variation between kerbs, steps, channels, low walls, pond edges and so on, which, of course, need pen sizes at the lower end.

As one gathers confidence, draughting speed increases and generally all lines become thinner. This may help the clarity of the drawings, so long as the relative difference between the line thickness has been maintained. Sometimes this differential is lost, so it is advisable to review pen size, drawing technique and line thickness being used, and refer back to the broad principles outlined above.

Ruled v. freehand

Ruled lines are crisp, clear and precise but they can look too crisp and impersonal when preparing drawings for clients unused to reading technical drawings. For most client drawings a more user friendly, handdrawn line may be preferable. Indeed this principle applies to plans produced on computer-aided design software. (For large client companies or public authorities and for larger-sized schemes it can be useful to be able to provide 'walk-through' computer-generated three-dimensional images for a client, but until computer technology improves, the handdrawn line for plans, and most perspectives, is more user friendly, appropriate and often faster.)

Ruled lines are far slower to draw for the beginner as they can cause ink-spread under the rule edge and also take time to set up. The compensation for this is that the ruled line is always crisp and does not demand a steady hand. Although French curves (flexible rubber rulers) can be used for irregular curves, it takes time to bend them to the precise shape required. In the end, a steady hand is necessary in order to make drawing time and cost effective. A steady hand can in even the worst cases be learned with much, much practice. The difference in speed between using French curves and a freehand curve is very great and so it is recommended that students 'bite the bullet' at the start and persevere with freehand drawings as much as possible. Freehand lines, once mastered, give you the advantage of great confidence and flexibility, as well as speed. You will be able to make client sketches (in front of your client) and you will find that regular freehand drawing helps to develop and improve your architectural lettering style – important for some plan annotation even in this computer age.

To practise freehand drawing try starting with simple horizontal and vertical lines, then try some circles and free curves. After all, horizontal and vertical lines and curves are the main components of most technical drawings and of letters too. There are a few tricks that sometimes help to focus your mind on the lines you draw and improve their quality. When drawing freehand circles, think of yourself looking down on to the top of a coffee mug. Strange, but it works. If you find that it does not work for you then resort to the circle template. With horizontal lines think of a maritime horizon as you draw the line. For vertical lines allow your hand to slide naturally down the page under gravity, keeping your hand and arm as relaxed as possible. At the end of the day, although these tricks may help, there is no substitute for practice and as much casual doodling and sketching should be encouraged as possible. This can be most usefully undertaken whilst on the telephone, in meetings or watching television.

Textures and shading

Differentiate in a bold graphic style the difference between the large masses of structural planting and the more careful shaping of lower-growing, more ornamental varieties.

Hatching

Hatching is used for large, solid objects such as buildings. There are three main ways of hatching (or shading with lines) an object. You can hatch to the right and then, to make another object stand out, you can hatch to the left. If you use both types on one plan it is advisable to hatch one type more densely than the other. The third method is cross-hatching where you hatch first to the right and then to the left to form a grid pattern. This will appear denser than the previous methods. The denser the hatching, the taller the effect. Always use a fine pen such as a 0.18 or 0.2.

Dotting

The use of dots is most applicable for uniform surfaces such as grass, gravel or tarmac. Enhance edges by making dots more dense and then gradually decrease the density towards the centre of the surface. Differentiate surfaces by altering both the size and density of dots. Generally dots should be made with the finer pen sizes, say from 0.18 (for grass) to 0.35 (sparsely dotted for tarmac or gravel).

Dense parallel lines

Closely spaced or dense parallel lines can make effective representation of bricks, setts and blocks but they must be drawn using a very fine pen.

Flecks and dotted parallel lines

If very fine pen sizes are used to make flecks (which can be all in the same direction or criss-crossed) or parallel dotted lines, then many soft or native surfaces can be shown, such as rough grass, wet marsh, native grass meadow or wild flower meadow. These textures can be used to fill in shrub symbols to differentiate between ornamental and native planting areas. Again, this is done with simple and bold graphics using coloured felt pens. Enclosing elements, including major planting masses, can be hatched in to define the spaces and pedestrian and vehicular routes can be drawn provisionally (using diagram graphics – dotted lines and arrows).

Colour

Colour is most often needed when preparing drawings for a client or for planning officers when planning permission is required for a development. Under these circumstances a presentation drawing is necessary in order to sell the scheme. Coloured plans stand out and show the effect of a proposed development much more clearly, particularly to the public or to the layperson unused to looking at plans. Shadows can be drawn on with dark colours to create a striking three-dimensional effect.

It is important for students to realize that colour should never be added to original tracing paper negatives. Designers use ink on tracing paper because the ink is very black and tracing paper is clear, allowing accurate prints to be taken from the plan and copies sent (at low cost) to the client, contractors, planners or anyone else involved. For enhancing a three-dimensional effect, crayon on a dyeline print is the best form. If the tracing paper plan (the negative) has any colour on it then the colours will show up in varying shades of grey on a dyeline print (blue being the lightest and yellow being the darkest). Even where the printing process is by photocopier, the use of colour on such a plan is mostly unsuccessful. Only where adhesive-backed blue plastic is applied to the plan can the effect of colour be used to advantage in the printing process. If a copy negative is taken then coloured pens and crayons may be used on the copy and can look very powerful when the negative is mounted against white paper, or where colour photocopies are taken (up to A1) but these are very expensive. However, the use of coloured pens on copy negatives is difficult because the ink takes a long time to dry and the negative must be mounted on a white paper background before it can be displayed. The best means of obtaining colour photocopies is to copy a coloured original onto plain tracing paper as, unlike white paper, the tracing paper is non-reflective and will not distort the colours. Generally it is best to apply colour to a dyeline print or photocopy, which being true paper have some absorbency unless the colouring is very complex and many colour copies are required.

Colour pens

Colour pens (especially large marker pens) have the advantage of allowing the rapid application of flat blocks of colour and are an excellent medium for diagrammatic or schematic drawings. However, these pens are by no means subtle and are inappropriate for delicate and naturalistic colouring work where many subtle shades of colour and tone are required to convey a true effect.

Crayons

Crayons, or even watercolour crayons, should never be overlooked in preference to marker pens because they allow more subtle mixing of colours and a broad range of textures.

Watercolours

The most luminous and subtle technique of all, with the greatest range of colour and tone, is the use of watercolour paint, though this is one of the hardest materials to use well and requires much skill and not a little technique. The use of tone is paramount with watercolour, that is, the relative lightness and darkness of the pigment. With watercolour you are not so much painting the features as painting their shadows to highlight them. With careful colour choice of the main washes and delicate drawing of the dark tones with the brush, quite stunning plans can be prepared with great depth.

Colour choice

Whatever medium is chosen, there are a few important points to remember when using colour. Warm colours such as reds and yellows have the effect of advancing towards you, cold colours such as blues and greens recede from you. This knowledge is most useful when colouring perspective renditions, when the added depth that the correct use of colour can create can be enhanced by manipulating the amount of contrast between light and dark. A great contrast between light and dark will appear to advance towards the viewer, while bland uniform textures (with little contrast between light and dark) appear to recede away from you. Indeed when drawing water (in perspective rendition) it is important to note that the reflections in the water of trees and other items on the bank are always a few tones darker, the contrast reduced and the images slightly fuzzy.

When colouring surfaces on a plan or in perspective renditions, a useful tip is to use darker tones to emphasize edges, in the same way as you would use dots and textures with your ink pen.

It is possible to create depth when colouring plans by toning down items and surfaces near to the ground by using light applications of complementary colours. (Complementary colours are those 'opposite' to the chosen colour for an object. For blue, the complementary colour is the sum of the other primary colours, i.e. yellow and red = orange.) By almost imperceptibly over-shading with a complementary colour you will tone down the main colour and make it recede away from you. This will have the dual effect of making other surrounding colours, on taller objects (such as shrubs, trees, buildings and walls), advance towards you.

Colours should be chosen to fit the materials shown on the plan. Trees and shrubs will obviously be shown in varying shades of green but gravel surfaces will require a fawn brown, while brick walls need a bright red to make them show up on plan and brick paving is best shown using a red-brown colour. Slabs and setts require a warm grey as cold or slate greys and dark slate blues are ideal for shading in shadows.

Shadows and tone

Shadows should be drawn to reflect the shape of the object casting them and the direction of the shadow should be consistent throughout the plan, from north-west to north-east. The precise angle should be determined by the orientation of the principal objects shown on the plan. It is always best to draw the shadows at an angle to major objects such as buildings, so that a shadow is cast on two sides of a rectangular building.

Shadows should be drawn with some direction in the stroke; that is, the stroke of the brush or crayon must be consistent in a direction that should be diametrically opposed to the direction of the light source. The shadow of a tree may be shown as a blob but this blob will be a little detached from the green circle of the tree (in plan) because of the height of the tree's stem, which itself casts a shadow.

The strength of the dark pigment (usually slate blue) used for shadows can sometimes dominate the plan and the use of the complementary colour (usually orange) will tone down the shadow accordingly.

When applying tone to a drawing (tone is the lightness and darkness of the pigment and has nothing to do with colour – a shade of, say, yellow can be made lighter or darker in tone) it is important to first determine from which direction the light source will come so that you can ensure dark tones for the shaded side and lighter tones for the sunny side of such objects as trees. These dark toned shaded sides of objects must be consistent in direction to appear convincing. When colouring perspective renditions and drawing shadows cast on to vertical surfaces the colour of the shadow will need to be warmer than a shadow cast on the ground.

Symbols

Symbols on plans

In order to represent any item, structure or surface that a designer might wish to place in a landscape, the draughtsperson must use a range of symbols. These symbols need to identify and distinguish clearly between all the different elements on the plan. Such elements will include incidental items (such as seats, lighting columns, litter bins) and singular artefacts (bird baths, sundials, sculptures) as well as more general items such as shrub planting, grass or paving (see Figures 15.4 and 15.5).

Not all these items will need to be shown in the key (sometimes called a legend). Some symbols may be useful and a key is always beneficial, but be selective. Do not waste time adding an item to a key when there is only one of the item on the plan. Any item that repeats several times, is a general item or extensively used surfacing will certainly need to be shown in the key. However, any singular items can be labelled – usually using a 0.2 pointer taken either vertically or horizontally (only diagonally if absolutely necessary, and then ensure the angles are consistent) to the side of the drawing. Pointers should end with a 1 mm round blob. If there is plenty of room on the plan and the lettering is computer generated, transfer or particularly well written by hand, then the label

Figure 15.4
Suggested symbols for commonly used items

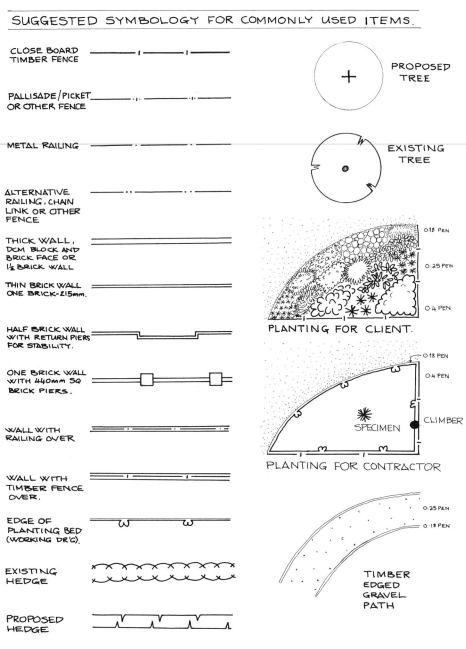

can be shown on the plan. If not, ensure the lettering is positioned well out to the side of the drawing. Unless it meets these criteria the lettering will detract from the drawing. For all but such singular items the key saves much repeat labelling and confusion over use of pointers.

Symbols on drawings for clients

The symbols chosen must distinguish between existing and proposed items, including shrubs and trees. Some of the symbols chosen will be different on a design drawing for a client from those chosen for a working drawing, prepared to inform a contractor how to build the scheme. Such symbols and graphics will be expressive of the subject drawn and will reveal some of the individual graphic style of each designer. Accuracy is important but generally takes second place to graphic style and effect. Such effects have to convey the ideas, the essential character that is given to the site by the design; that is, the graphics sell the client the designer's dream.

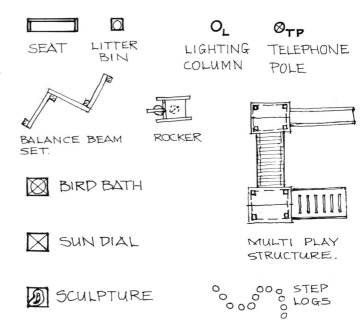

FIGURE 15.5 SYMBOLS. Whilst there are British Standard line types for different fences, planting and other surfaces and materials there are many artefacts and features that do not have conventional symbols, and the above may provide some assistance in draughting such items.

When preparing drawings for a client the symbols used for soft works items are intended to indicate the effects proposed by the designer but without any detailed information provided about specific plants. The graphics should show the relative height of the plants, perhaps in three bands – low, medium and tall (if the bed is wide enough). The symbols will also show the textures in three types – fine, medium and coarse. The habit of the plant can also be depicted to some extent by choosing a predetermined range of symbols to show a variety of plant character, both in habit and texture. Height and character can be illustrated by variation in the pen thickness and the pen stroke respectively, using pens from 0.2 to 0.6. Trees can be drawn showing winter twigs as radial, slightly swirling fronds using a fine pen or in summer garb using leaf shapes or irregular indents, but still following a true circle outline.

Street furniture is not easy to show in plan, but must be drawn as evocatively as possible – using lines to show the characteristics of the item. For a seat, one would show the back, armrests and possibly the slats (in a finer pen). Most street furniture is depicted with circular, square or rectangular shapes, drawn to scale and shaded, crossed, blocked in or not – to suit.

Symbols for contractors' working drawings

The symbols for design drawings tend to be personal and reflect the individual style of the draughtsperson, but for working drawings the symbols must possess simplicity, clarity, uniformity and consistency – so that the design can be easily understood by those carrying out the works on site. The working drawings must conform to convention and the symbols used must be nationally, even internationally, recognizable. The United Kingdom has a British Standard code for such symbols (BS 1192: 1984 Part 4). In practice, however, the unwritten code prevails and a general (rather than a fundamentalist) adherence to the convention is sufficient.

Symbols for enclosing elements

Enclosing elements include timber, wire and metal fencing; railings; hedges; walls of stone, blocks or bricks; trellis work, or tall shrubs. Some enclosure is implied by low level items such as steps and kerbs, or even intermittent items like bollards or trees.

All these items can be shown by using lines, dotted lines, spots and dashes as explained earlier, but although it is best to adhere to the conventions there are necessarily many variations of the standard symbols and clarity is the most important factor. As long as the symbol chosen reminds the viewer of what it tries to portray, is not a symbol normally used for another purpose and is clear and shown in the key or legend, it will be acceptable. The problem with ignoring the conventional symbols is that the plan may cause confusion and be misread in haste.

A broad affiliation with the conventional symbols as set out in BS 1192: 1984 Part 4, but with sufficient variations to be able to show many types of fence (for example) on the same plan, will generally be the best approach. This applies equally to all drawings when concerned with enclosing elements, with the important exception of shrub planting on design drawings for a client. For a design drawing the British Standard symbol would be dull, unimpressive and inappropriate; more suggestive graphics are essential for such drawings. Some useful symbols are set out below:

1. Railings – dash-dot-dash or dash-dot-dot-dash if a second or more types of railing are used; 0.6 pen.
2. Walls – two parallel lines close together; 0.4 pen.
3. Fencing over a wall – as above but with dash-dot-dash in the middle; 0.4 pen.
4. Close board timber fencing – dash-blip-dash, the blip being a short line about 2 mm long at right angles to the dash, but not touching either side; 0.6 pen.
5. Palisade timber fencing – dash-dot-blip-dot-dash; usually 0.6 pen.
6. A further timber fence, such as a picket fence – dash-dash-blip-dash-dash; 0.6 pen is usual.
7. Post and wire fences and chain-link fencing – dash-dot-dot-dot-dash or similar, 0.6 pen.
8. Steps – 0.4 pen for the risers, with an arrow positioned centrally – using a 0.18 pen with the arrow's head pointing up the steps; side walls are also shown in a 0.4 pen, as a double line.
9. Kerbs – 0.25 pen for the forward edge; 0.18 pen for the back edge.

10. Bollards – 0.4 pen to create a small blob or spot or even a square; joined up with a 0.4 line to represent a knee-rail.

All the dashes referred to should be around 2 to 2.5 cm long, generally speaking, and while the depiction of walls will demand no more than a 0.4 pen to give them sufficient presence, both fence and railing lines will require a 0.6, 0.7 or even 0.8 pen to show up clearly, being of single line thickness only. Even a high wall of, say, 2 m will normally need no more than a 0.4 pen because of the perceptual effect of two parallel lines (as discussed above).

Symbols for soft works

Soft items such as shrubs, hedges and trees can be represented in an abstracted clinical way for working drawings, but on design drawings (for a client) great variety of texture and graphics style can be employed – personal to the individual designer. Trees may be shown with leaf symbols or branches, often with denser patterning to the shaded side. Shrubs will be portrayed using a variety of swirls, stars, blobs and serrated whorls, with a 0.25 pen for ground cover plants at the front of borders, a 0.4 pen for medium height shrubs in the middle of borders and a 0.6 pen to show structural planting at the back of a bed.

For working drawings, including the planting plan, the story is quite different. Once again there are conventional symbols for all soft works items, as set out by the British Standards Institute (BS 1192: 1984 Part 4) along with all the others listed in this book. In fact a total of 83 symbols for gardening, horticulture and landscape work can be obtained from the BSI at Milton Keynes – the sectional list SL41, which sets all these out, can be obtained free of charge.

Proposed hedges on working drawings are usually depicted as a double line (showing each side of the hedge), each line punctuated every 2.5 cm with a narrow 'V'. Existing hedges are normally shown as a series of interlocking shallow semicircles – 1–2 cm long for each side of the hedge. Proposed trees are most commonly shown as a simple cross in a 0.6 pen (despite what the British Standard says) with a circle drawn in a 0.18 pen – approximately measuring 4 m in diameter – to the scale of the drawing. Existing tree stems are shown as circles in a 0.4 or 0.6 pen, and hatched with a 0.18 pen. The canopy is drawn to the shape that it exists on site as a series of arcs in a 0.4 pen, punctuated with long narrow 'W'-type indentations.

Proposed shrub or herbaceous planting is shown as a 0.4 line punctuated with an indented lower case 'w' every 2–2.5 cm, one time affectionately referred to as 'babies' bums' by a student, a phrase that has stuck ever since. Because this symbol represents the height of the actual planting, the bed edge needs defining outside the planting line and this is achieved by a straight line about 1 mm from the planting symbol, using a 0.18 pen. The planting symbol is drawn right around the bed to show the extent of the planting and for planting plans the planting area is then subdivided into blocks (using a 0.2 size pen) which are then labelled. Usually the labelling of each block is achieved by pointers (0.18 pen) leading out to the side of the bed where the plant quantity and name are written. This 'block planting' method is not the only method but is arguably the fastest and certainly the most commonly used.

The standard planting symbol for working drawings will be used to show the location of planted areas for both the hard works layout plan and the setting out plan, where no subdivision will be necessary.

Existing planting is generally depicted by using a continuous small lobed line, like a continuous joined up lower case 'm', right around the planting area. This planting, like the trees, can be numbered for inclusion on a schedule of felling and surgery works.

Summary of drawing types

Design drawings

Textures – Yes
Shading – Yes
Dotting – Yes
Hatching – Yes
Symbols – Yes
Colouring – Yes
Conventional? – Not necessary
Pretty – Yes
Notes: For the client. These should be brief without detailed specification, but should be descriptive.

Working drawings

Layout plan
Textures – Yes
Shading – Yes
Dotting – Yes
Hatching – Yes
Symbols – Yes
Colouring – No
Conventional? – Yes
Pretty – No
Notes: For the contractor. They should be written as commands and be concise and clear, specifying quality and quantity of workmanship and materials.

Setting out plan
Textures – No
Shading – No
Dotting – No
Hatching – No

Symbols – No
Colouring – No
Conventional? – Yes
Pretty – No
Notes: For the contractor. They should be brief and concise.

Demolition plan
Textures – Yes
Shading – Yes
Dotting – Yes
Hatching – Yes
Symbols – Yes
Colouring – No
Conventional? – Yes
Pretty – No
Notes: For the contractor. They should be brief and concise.

Construction details
Textures – No
Shading – No
Dotting – Yes
Hatching – Yes
Symbols – No
Colouring – No
Conventional? Yes
Pretty – No
Notes: Detailed specification for construction purposes.

Other drawing types to aid communication

Perspective rendition

The principal purpose of perspective rendition in landscape design is for the presentation of design ideas for the client. Although the age of computer-generated 3D image is here, with walk-through facility and sophisticated photomontage which can give dramatic effects, nothing quite compares to a well-drawn perspective rendition. The rendition has one supreme advantage over the computer image, it possesses 'soul'. At the end of the day we are designing landscapes to improve our world and we are improving it mostly to satisfy functions in a more aesthetically pleasing manner, for the good of all our souls if you will. Arguably the best media for this task is one that best reflects the character, the mood and the soul of the place, the designer and indeed the users too.

Before you commence a sketch it is necessary to decide on the eye level. This is the term given to the height of the viewer above the ground. A high eye level, perhaps the vantage of someone standing on a chair, will put more emphasis on the land than on the sky, while a low eye level will place more emphasis on the sky. The watercolourist John Sell Cotman used to manipulate eye level to the full with some of his paintings, the land occupying one-fifth of the paper, while the sky filled the rest and became the main focus, with dramatic cloud effects across flat, fenland landscapes. In reverse, where it is important to show the detail of the landscape itself, such as that of a busy street, artists will choose the eye level of someone on the first or second floor of a building, and then the sky might occupy one-fifth of the paper. For landscape design perspective rendition this high eye level is always preferable, to show up the detail of the surface treatments and to avoid any enclosing elements obscuring too much of the view.

Having determined a satisfactory eye level, the vanishing point must be chosen. There might be more than one of these; that is, there might be more than one vista. The vanishing point is the point on the horizon where all lines converge and the eye can see no further. If this point is to the right or left of the centre line of the page, then it likely that there may be a second vista in another direction. A single vanishing point is called single-point perspective. Where there are two vanishing points it is called two-point perspective.

Starting with single-point perspective, choose the position for the vanishing point along the horizon line (the eye level), to determine the direction of the view. If the vanishing point is moved to the right, then the view will look across the page to the right and so on.

It is worth mentioning that the vanishing point does not have to be exactly on the horizon line, unless the ground is flat. If you place the vanishing point above the horizon line then the picture will appear to look uphill. Though it is true that the vanishing point cannot be below the horizon line (unless the picture were looking down a hole), you can make the picture appear to be looking downhill by aiming all lines travelling away from the viewer to a secondary point below the horizon line. These lines will have to return to the horizon line at some point, which will be the point that the slope levels out.

Once you have chosen the position of the vanishing point you can draw some faint lines out to the sides of the paper at regular intervals as guide lines. These are the convergence lines and any line in the picture that is travelling away from the viewer will travel along this convergence line. This means that as two parallel lines (say of a path) travel away from the viewer the lines will become closer together until they eventually meet. The reducing size of objects moved away from the viewer is called foreshortening. However, any line travelling at the same distance from the viewer will be drawn absolutely horizontal. Any line that is rising vertically will be absolutely vertical.

Near objects will demand more detail than those further from the viewer, which can be drawn in a far more suggestive way; the eye will interpret the suggestion as well as a finely drawn object, if not better, though the near objects will be studied more closely. Tone, or the lightness and darkness of an object, may be added to the line rendition to add solidity and reinforce the sense of plane. This will greatly enhance the feeling of depth.

Two-point perspective will require two vanishing points at each end of the page. Any line that is travelling away from the viewer will travel to one or other of the convergence lines. Lines that are perpendicular to one convergence line will travel towards the other.

Three-point perspective will add a third vanishing point, which will most often be a vertical one. That is to say, as the vertical lines travel up and away from the viewer, they too converge at a point in the sky and so buildings, trees, lamp columns and so on will have their vertical lines slanting towards the third vanishing point and therefore such objects will taper and become thinner the higher up they are.

Axonometric projection

There are occasions where an accurate three-dimensional rendition is required, not from the vantage of the pedestrian but from the vantage of someone standing on a roof top, in order to reveal the whole design and in a proportionate, measured way, to scale. If perspective rendition were used then the foreshortening would mean that only some areas of the design would be explained well while others would be hidden. Therefore, where an overall picture of a design is required, and especially where the client or user group are not used to looking at plans, axonometric projection is essential.

A straightforward design plan is tilted at an angle of 45 degrees from the horizontal and then all items that have any height to them have vertical lines drawn up to their full height – measured using the same scale as the plan. Then the tops of the objects are drawn to the same width and depth as the plan view. With soft works the vertical lines are embellished with textures to mimic the leaves of trees.

Isometric projection

An isometric projection is very similar to the axonometric but the plan that is to be projected will be tilted at 30 degrees to the vertical instead of 45 degrees, which produces the effect of a slightly lower vantage point.

Scaling up and down

Landscape designers will continually find that they need to scale plans/drawings up or down in order to meet the client's requirements.

Manual (or grid) method

This method is not often used in this day and age of the super copier, but it is useful to know about. Draw a grid over the plan at 5 m intervals. Draw a grid at the reduced or enlarged scale in pencil with the lines 5 m apart (using the new scale). Number the boxes down each edge. Draught lines of the plan in accordance with the relevant boxes, then ink in on the back of the tracing paper, following the pencil lines, and finally rub out the pencil grid on the other side.

Photocopier method

Modern copiers are very clear and accurate. Blow up (up to 200 per cent in one go) and use the scale rule to check for accuracy and distortion. Choose a known feature on the original plan and, once enlarged on the machine, check the same feature at the intended new scale against the old plan at the old scale. Adjust the enlargement percentage if necessary. Reverse if scaling down. All copying causes some distortion, though this can be mitigated by adjusting the position of the plan and copying twice instead of once. If a distortion occurs in one direction, copy again having first rotated the plan 90 degrees, ensuring that the enlargement is equally split between each copying process.

Table of photocopying

Enlargements

1:500	to 1:100	$= 1 \times 200\% + 1 \times 125\% + 1 \times 200\%$
1:500	to 1:200	$= 1 \times 200\% + 1 \times 125\%$
1:500	to 1:250	$= 1 \times 200\%$
1:500	to 1:50	$= 1 \times 200\% + 1 \times 125\% + 1 \times 200\% + 1 \times 200\%$
1:1250	to 1:500	$= 1 \times 200\% + 1 \times 125\%$
1:200	to 1:100	$= 1 \times 200\%$
1:200	to 1:50	$= 1 \times 200\% + 1 \times 200\%$

Reductions

1:100	to 1:500	$= 1 \times 50\% + 1 \times 75\% + 1 \times 50\%$
1:200	to 1:500	$= 1 \times 75\% + 1 \times 50\%$
1:250	to 1:500	$= 1 \times 50\%$
1:50	to 1:500	$= 1 \times 50\% + 1 \times 75\% + 1 \times 50\% + 1 \times 50\%$
1:500	to 1:1250	$= 1 \times 50\% + 1 \times 80\%$
1:200	to 1:100	$= 1 \times 50\%$
1:50	to 1:200	$= 1 \times 50\% + 1 \times 50\%$

CHAPTER 16

Drawing arrangement

BEFORE commencing any drawing, whether a design drawing for a client or a working drawing for a contractor, it is important to take time to consider how much additional text information is going to be located around the pen drawing; to determine at what scale the plan must be drawn to show clearly what is trying to be portrayed; and to consider how all the information will be arranged on the sheet. There is nothing worse than spending hours preparing a drawing, only to find that there is too much information to fit on one sheet and that it must all be condensed to a point at which it becomes unreadable or hopelessly cluttered. It is always better to have too much space than too little. Excessive space can be made to look acceptable by carefully arranging the information on the sheet, but will look absurd if all the information is packed tightly into one corner, leaving large areas of the remaining sheet blank. Some thought must also be given to the sequence in which the information should be read. Clearly, somebody unfamiliar with the drawing will first need to look at the title block to determine what the drawing is about, which site it belongs to and so on, even before the plan is unfolded. It is for this reason that there are explicit conventions for the placing of common elements to all drawings (such as the title block being placed in the bottom right-hand corner of the sheet).

Scale and paper size

Judging the size of the paper to use for a plan is very important to the attractiveness of the plan and presentation of the information on it. It looks equally bad if the plan is too cramped or too spaced out. However, when choosing a scale for plans of any kind, it is necessary to match the size of drawing to the paper size, allowing plenty of spare room for the key, title block, notes, north sign and location plan.

The paper sizes should be those of standard size, such as A0, A1, A2, A3 and A4, the last two sizes mostly being used for construction details. The standard paper sizes will allow ease of handling and ensure that the tracing paper drawings (negatives) can be printed easily without missing off information.

The notes on a plan can take up either a little room, in the case of design drawings, or a large amount of space, in the case of working drawings such as the planting plan and layout plans. It is therefore always better to choose a paper size that at first sight is too big, rather than choose one that ends up too small for all the information eventually required on it. There is nothing worse for the contractor on site than to have many A4 attachments, paper-clipped to the plan, all falling off at once on a windy day.

Estimate the amount of information likely to be required in the notes, determine the number of other items to be added to the plan and work out how much useful space you will have on the plan after the site area itself has been drawn. The site orientation is vital in this respect. Convention dictates that north should always be pointing up the paper but in reality this is not always practical and it is more important to fit everything on to a standard size sheet of paper than to align the site to convention.

If you first estimate the area likely to be taken up by notes, key, north sign and title block you can work out their arrangement around the drawing by cutting out a piece of paper to fit the size of each item. Label each piece of paper and place the paper items around the drawing in the most practical arrangement (the title block should wherever possible be positioned in the bottom right-hand corner). Try to make the information flow in a logical sequence. Choose the paper size accordingly.

Appropriate scales for typical drawings

Location plans: 1: 500–1: 2500. For client and contractor.
Concept and masterplans: 1: 1250–1: 500 (small site 1: 200). For client.
Design plans – hard and soft: 1: 200–1: 100. For client.
Axonometric drawing: 1: 200–1: 50. For client.
Perspective renditions: 1: 200–1: 100 (often freehand). For client.
Layout plans – hard and soft: 1: 200–1: 100. For contractor.
Setting out plan: 1: 200–1: 100. For contractor.
Demolition plan: 1: 200–1: 100. For contractor.
Planting plan: 1: 200–1: 50. For contractor.
Detail layout plans: 1: 50. For contractor.
Construction details: 1: 50–1: 5 (usually 1: 10 or 1: 20). For contractor.

Common elements of drawings

Title block

The title block is a ruled box with several compartments which is filled in by the draughtsperson to inform its users of the drawing's content and purpose. The title block is always placed in the bottom right-hand corner, so that when the plans are folded it will be clearly seen on the front. This is a convention well founded in practical good sense as it ensures that the plan can be easily found and retrieved from among many other related drawings, even long after it has been forgotten about and filed away. The title block should sport your own (or company) logo and below it should be set out a series of boxes for the following items – in order of appearance:

1. The client name.
2. The site name.
3. The drawing title.
4. The scale.
5. The date.
6. The initials of the draughtsperson.
7. The initials of the person(s) responsible for checking the work.
8. The job and drawing no. plus any revisions A, B, C, etc.

The key (or legend)

A key comprises a series of boxes lined up beneath the word KEY (or LEGEND), itself written in large bold letters so that it is easily noticed. The boxes must be large enough to show the items in them clearly and a box 1.5 × 2.5 cm is ideal. As noted earlier, only items that repeat several times need be put into the key; isolated items should be labelled. Ensure that the boxes are spaced both for clarity and to provide enough room for the explanation. Each explanation should be written in architectural lettering (usually upper case) or computer lettering. The wording should be concise but clear and thorough so that the reader is in no doubt as to the content. It is always visually more attractive to space the boxes evenly, but this is less often possible on working drawings. Fortunately it is less important for such plans to be aesthetically pleasing.

All singular artefacts or features on the drawing should be labelled, rather than overcomplicate the key. It is also quicker to read a label (so long as there are not too many of them) than it is to refer to a key away from the drawing itself.

Notes

Notes may be for the client or for the contractor. Client notes need to be simple, concisely descriptive and clear. Contractors' notes need to be precise, thorough but not verbose, and phrased as specific commands such as 'Supply and construct 38 × 125 mm timber edging board . . . '

Keep all notes away from the plan to avoid clutter, utilizing pointers – horizontal or vertical in the main for neatness but diagonals if absolutely necessary.

When annotating working drawings it is essential not to duplicate information, to avoid both wasting time and risking confusion or discrepancies between plans. If a construction detail for an element on the plan is needed, there is no need to duplicate the information on both plan and the detail. The detail should simply be referred to on the layout plan, for example 'Supply and construct concrete slab paving as specified on detail drawing D1.' The key should be equally explicit.

There should be references to construction details on the layout plan, such as a number (D1) in a circle and pointer to the item or (less satisfactorily) just a pointer to the item, described at the side of the actual plan. Levels should be shown (existing and proposed), proposed levels in a box. Levels on walls should be shown with a bar over. Manholes, drainage runs, service runs, gulleys and falls should also be shown.

The north sign

These are essential for landscape designers, to determine the amount of light and shade and the specific microclimatic conditions within different parts of the site. North signs are best if kept very simple. Place them out of the way, but where they can be seen clearly. Make sure they are orientated as accurately as possible – essential for preparing planting plans.

The location plan

Location plans are always useful to help place the site in the wider context of the surrounding area and to help subconsultants and contractors find the site quickly and easily. They are almost always drawn at large scale (e.g. 1:1250 or 1:2500) with key roads, access points and major landmarks labelled and with their own north sign.

Layout on paper

It is important to arrange the information using the conventional format of title block in bottom right-hand corner, key above it and north sign above key if possible. Other spaces can contain the location plan and notes, which should be set out in the most logical manner in 'newspaper' (readable) columns with the information flowing from left to right.

Computers

It is important to understand the enormous benefits of computers and some possible pitfalls, but worth pointing out that the speed of advance of computer technology is so fast that by the time this book is published technology will have developed further.

Word processing

Word processing is arguably still the most important function of the computer for landscape work. Standard forms, letters, invoices, schedules of quantities, cost estimates, reports, appraisal documents, schedules of plants, tree surveys and contract

administration forms can all be produced on word processing and spreadsheet software. The schedules of plants and tree surveys can be printed on to paper and
photocopied on to self-adhesive film which can then be stuck on to drawing negatives. The time saving is extraordinary when one thinks back to the days of handwriting everything and then requesting your secretary type it up from scratch each
and every time. The ability to copy old files into new files has revolutionized the
whole process of administration, documentation and document retrieval.

The software available for landscape design is quite varied including
sophisticated computer-aided design packages which the designer can use to
prepare most aspects of landscape design including planting plans, ground
modelling schemes and construction details. There are also useful CD ROMs
available such as from the RHS on poisonous plants.

Computer-aided design

Advances in computer-aided design (CAD) have been dramatic in recent years,
and disks are increasingly being sent to clients through the post, or relayed
through the Internet, far more easily than tubes containing precious plans pre-
cariously being sent parcel post. Landscape details can be added to layout plans
much faster than by handdrawing on to a tracing paper negative. However,
many people still prefer the handdrawn plan, especially for design drawings,
and it should not be forgotten that the human brain is the best computer of all,
especially when it comes to choosing plants or imagining design solutions.

Having said that, there is no getting away from the advantages of CAD.
The use of the pull-down menu to allow plant selection simply with the click of
a mouse can speed up the design of planting schemes massively, at least for the
simpler development schemes. The computer can be programmed to calculate
the area of the bed, multiply this area by a predetermined plant spacing and
automatically print the plant name and quantity on to the plan with a pointer to
the section of the bed chosen. Even more useful is the way the computer trans-
fers this information to a plant schedule and prints the completed plant schedule on to the drawing.

When evaluating whether to invest in CAD it is worth noting just what
the real time saving advances are. The practised human landscape designer can
choose a plant in a few seconds, measure the area in a few seconds too, calculate the number of plants required by multiplying the spacing by the area in just
a few seconds more, and the skilled practitioner will have written the quantity
and plant name very shortly afterwards. It is unlikely that the computer will be
able to perform this task very much faster, but the automatic 'take off' (measuring from a plan) of quantities and plant schedules and the computer's ability to
check for mistakes have indeed been revolutionary. It is this stage, which in the
past would have taken even the most skilled practitioner hours of laborious
work, that can now be carried out in the twinkling of the designer's eye.

Computer-aided design has revolutionized the draughting of working
drawings and details, at least in the hands of a skilled operative. CAD is the
future but there will always be a place for the handdrawn plan – the nature of

our profession is about the whimsical, about people's leisure, about aesthetics as well as about utility, and the handdrawing conveys the human touch so much better than the machine and is more appropriate for the design drawings at least. For how much longer this will be the case it is difficult to say.

Lettering for drawings

As explained earlier, architectural lettering is useful for handwriting text on to a drawing in a clear and precise manner, which can be easily read by any professional contractor or interested party. The lettering must therefore be to established lettering conventions so that readers are not faced with scratchy and illegible handwriting. Having said that, there is an amazing amount of variation in the styling of architectural lettering but, above all, clarity is the principal requirement.

Upper and lower case

Upper case is the term given to wording written in capital letters; lower case applies to small letters. Most annotation on plans is written using upper case letters. This annotation is referred to as architectural lettering, which is most often handwritten in a clear and printed style. Every designer will have a different style of architectural lettering, though the amount seen is ever-decreasing as computers allow faster annotation of plans, the text being transferred by proprietary photocopiable self-adhesive film. Indeed, with computer-aided design both the drawing and its attendant lettering are plotted directly on the paper.

There are a few golden rules with lettering. Never mix size, style or case in any one line. This is especially true for labelling and titles. Notes can be written using lower case, but upper case is always clearer and therefore is preferable.

Clarity and style

Variety in style helps to provide both personal or corporate identity, but it is essential that the script is clear. Old English, for example, is almost illegible, especially at a distance, whereas Goudy Bold is traditional, attractive and clear. Helvetica Bold is also clear, but rather pedestrian and dull.

It is possible to create your own individual style of architectural lettering by exaggerating one letter in the alphabet to give rhythm, for example enlarge the 'O' or put a tail on the 'R' or play around with the 'E' and so on. Above all, lettering must be clear, sufficiently large to be read at a glance from a short distance and succinct.

Techniques to improve neatness and style

It is always advisable to draught letters between two parallel pencil lines or use a metric graph paper sheet under the tracing paper negative as a guide. Elongate letters for added neatness and take care with A, M, W and Y. The horizontal line of A should be below the half way line, while O and Q should extend very slightly above the top and bottom guide line in order not to appear smaller than other letters (this is another of those strange optical effects which it is important for the designer to understand).

The spacing between letters is more important for neatness than the letters themselves. Try to make the spaces as even in size as possible (take care with the letter I and the juxtaposition of any letter with vertical sides, particularly when these are next to A and L, which have large spaces around them). The key is to even out the size of the space between the letters rather than evenly space the letters themselves.

Freehand, stencil and computer lettering

Using a letter stencil is very slow, and though a stencil can produce very neat lettering in skilled hands, this neatness only comes with much practice. The use of a stencil is not commercially viable for most landscape commissions, freehand letters being much faster and in the correct hands almost as neat. Freehand lettering is far more user friendly, too, which is important for clients. It also adds that individual touch of personal style.

Fastest and neatest of all is to use a computer and print your notes out and photocopy them on to Transtext or other proprietary transfer film. The lettering is rather lacking in personality, but at least, using the spellchecker, the spelling will be reliable. Computer lettering is ideal for all working drawings.

Letter hierarchy

When setting out information pay attention to the letter hierarchy in order to convey visually the relative importance of the titles and headings. This hierarchy will display how the information you are presenting all fits together. As a general guide use the following hierarchy:

1. Upper case bold and underlined.
2. Upper case bold.
3. Upper case underlined.
4. Upper case.
5. Lower case.

It is contrary to convention to underline lettering when using freehand architectural lettering, when the size of the letters and the spacing of lines is generally thought to be sufficient to convey hierarchy.

Lettering on a plan

The height of letters can be varied for different purposes. Use large letters for all public and client consultation drawings (3 mm plus) and smaller letters (2 mm) for working drawings – or use a computer typeface. Keep all labels (bar a very few, brief ones) off the plan, using pointers. Pointers are covered elsewhere, but it is worth noting that where there is much lettering and labelling, pointers should be drawn either horizontally or vertically unless it is absolutely necessary to use diagonal lines. If the need arises, these should at least be at a consistent angle.

CHAPTER 17

Drawing for different purposes

THERE are two primary functions of drawings: first, marketing design ideas to the client and, second, instructing the contractor how to build the design ideas.

Marketing ideas and marketing a landscape practice are directly or indirectly achieved through the practice drawings. Sloppy design and draughting will not help to sell a client the ideas (however well founded these ideas are) and above all will not help to sell the practice as competent and professional. Design drawings for the client must be particularly well drawn, both because they are selling the design to the client and because they are the first drawings that the client will see and are likely to be the ones best understood too.

The other main purpose of drawings is to instruct those who are going to build the landscape on how and where to build it. For this purpose the drawings must be precise and thorough, giving every detail of information required or referring to other drawings that provide this information. Indeed there will be a series of drawings required for every site – bar the simplest of schemes – each providing specific information, from general layout to close enlargements of components and how they fit together. Clearly, drawings meant for contractors must use familiar conventional graphics and be clear, simple and impersonal in style, providing only the information required at any one stage of the works.

Targeting information

The main intention of having different drawings for different purposes is to allow the contractor to be able to find just the information needed at the time and no more. If all the information required to build a landscape were on the same drawing, then clearly it would become confusing and it would also take time to find the relevant information.

When producing a set of drawings for a site it is essential that the information is not duplicated. Each drawing will have its own notes and information. The hard works layout plan should not have a full specification for, say, a brick wall if there is a full construction detail that provides all the necessary information. Having said that, it is also vital that the drawings are satisfactorily cross-referenced so that the information needed is easy to find. There should be reference to other drawings in the notes and on the key, and detail references should be shown on the plan numbered D1, D2, etc. These references should be drawn in a circle (0.18) with a pointer to the element concerned.

It is important to recognize the difference between working drawings, which are used to provide information for the contractor, and design drawings, which are used to provide information to the client. The latter will not give construction information, may suggest options and might give brief explanations for the location of features or choice of materials. Working drawings for the contractor are annotated entirely differently; they are written as a series of concise commands, which offer no choice or explanation but crisply deliver the essential construction information required.

Sections and elevations – for the client

A section for a client will usually be drawn for illustrative purposes, to show how the site will appear from a side (or end) view, but having taken an imaginary slice through the site, perhaps including below ground, and transecting any objects in the path of the section line. Such sections will generally be at scales from 1:500 to 1:20, but mostly 1:50 and 1:20. They will show the profile of the site and indicate changes of level at walls, kerbs, banks and so on.

The items on the section line (proposed or existing) will be drawn boldly with a thick line, and such items will indicate space enclosure and the scale of the spaces. Items drawn further away will be drawn in a fine pen so that they appear to recede as background items. This latter detailing is essential to give the section character and convey the true appearance of the landscape being designed. The road/path or pavement base courses can be indicatively shown, to contrast with the hatching of areas of soft works. Whatever the surface treatment, the essential component of a successful section is a bold section line, which should empty at least a 1.2 pen size.

Elevations for a client are actual side or end views of the site and do not transect objects. For example an elevation of a building will show the side or end wall, complete with windows, climbing plants, perhaps, and any object between the viewing point and the wall of the building. A section would actually cut through the building and show the floor levels and so on (see Figure 17.3).

Neither sections nor elevations show any foreshortening, which is convergence of lines towards the horizon, as would be shown with perspective renditions.

Sections and elevations – for the contractor

Sections and elevations for a contractor will follow the above rules but with a few essential differences. The first point is that the sections will be used for a different purpose – to explain how the various component elements fit together. For this reason the sections will need to be drawn at much larger scales, from 1:50 to 1:10 and mostly from 1:20 to 1:10. The scale must be sufficient to show how individual paving units, for example, are actually joined to the next unit or to edgings or other components. The section will be drawn without embellishment, often with no or minimal background items shown.

The items that transect the section line will be drawn with a more bold outline than items just behind the section line. Not all items are drawn as boldly as others, however. Units are drawn with the boldest outlines (0.4 or 0.5), while foundations, haunching and bedding and base materials will be drawn with finer outlines, say, 0.25.

Plans for the client

The location plan

The location plan is not so much a plan as a part of other plans. It is normally produced by copying an extract from a local map, though sometimes a not-to-scale sketch is drawn (see Figure 17.1). Generally a plan derived from a map is preferable, though some additional labelling is often necessary to highlight land-marks and key roads to assist with finding the site. It is necessary to provide a scale if the plan is drawn to scale and also a north sign.

Concept drawings

Concept drawings will utilize the information provided on the survey plan and a base plan will be draughted from the survey plan showing just the items that are intended for retention. This base plan can be copied several times, perhaps reduced, and can be used to analyse the site using diagrams. All the site appraisal information can be depicted in diagrammatic form on these zone plans, for example zones of damp and dry areas can be coloured or shaded, or exposed and sheltered parts, sunny areas and shaded, private and overlooked and so on (see Figure 17.2).

Once all these proposed zone plans have been superimposed a further overlay plan can be drawn, which will shape the spaces and make them into a more harmonious arrangement. From this early 'fag packet' sketch, as they are endearingly known, the first sketch design plan can be produced. It is at this stage that it may be helpful to prepare perspective renditions to illustrate the ideas in a more user friendly way. This may be particularly helpful if the ideas need to be sold to the client or where there are meetings with the local planning authority.

Final design layout

The final design layout will be at a larger scale than the sketch design. It may be

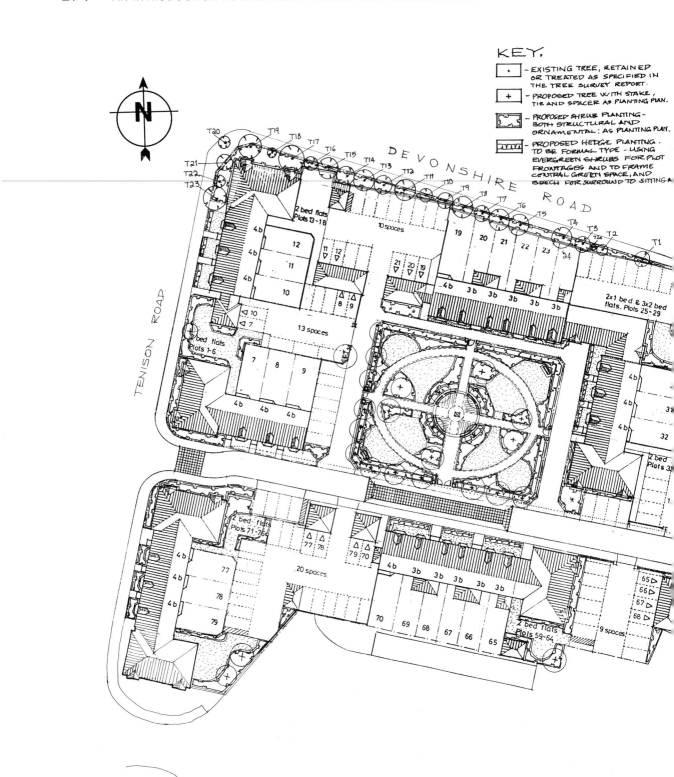

KEY.

☐• – EXISTING TREE, RETAINED OR TREATED AS SPECIFIED IN THE TREE SURVEY REPORT.

☐+ – PROPOSED TREE WITH STAKE, TIE AND SPACER AS PLANTING PLAN.

☐ – PROPOSED SHRUB PLANTING – BOTH STRUCTURAL AND ORNAMENTAL : AS PLANTING PLAN.

☐ – PROPOSED HEDGE PLANTING. TO BE FORMAL TYPE – USING EVERGREEN SHRUBS FOR PLOT FRONTAGES AND TO FRAME CENTRAL GREEN SPACE, AND BEECH FOR SURROUND TO SITTING A

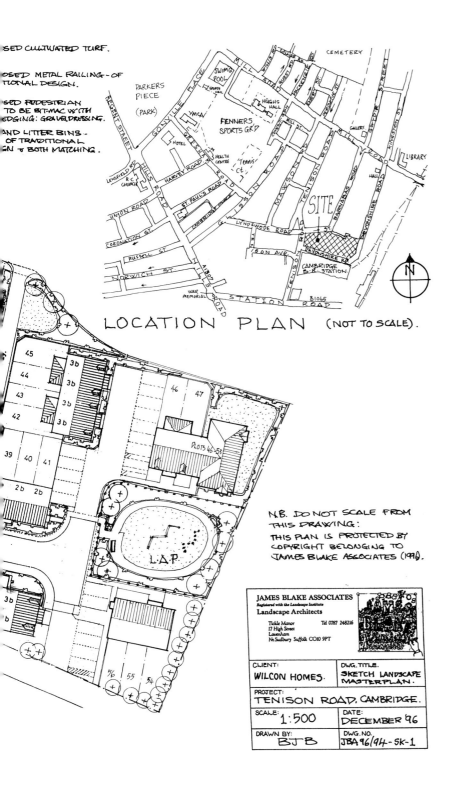

...SED CULTIVATED TURF.

...OSED METAL RAILING - OF
...TIONAL DESIGN.

...SED PEDESTRIAN
...TO BE BIT-MAC WITH
...EDGING: GRAVEL DRESSING.

...AND LITTER BINS -
...OF TRADITIONAL
...GN J BOTH MATCHING.

LOCATION PLAN (NOT TO SCALE).

N.B. DO NOT SCALE FROM
THIS DRAWING:
THIS PLAN IS PROTECTED BY
COPYRIGHT BELONGING TO
JAMES BLAKE ASSOCIATES (1996).

JAMES BLAKE ASSOCIATES
Registered with the Landscape Institute
Landscape Architects

Tickle Manor   Tel 0787 248216
17 High Street
Lavenham
Nr.Sudbury Suffolk CO10 9PT

CLIENT:	DWG. TITLE.
WILCON HOMES.	SKETCH LANDSCAPE MASTERPLAN.
PROJECT: TENISON ROAD. CAMBRIDGE.	
SCALE: 1:500	DATE: DECEMBER 96
DRAWN BY: BJB	DWG. NO. JBA 96/94-SK-1

FIGURE 17.1 LOCATION PLAN

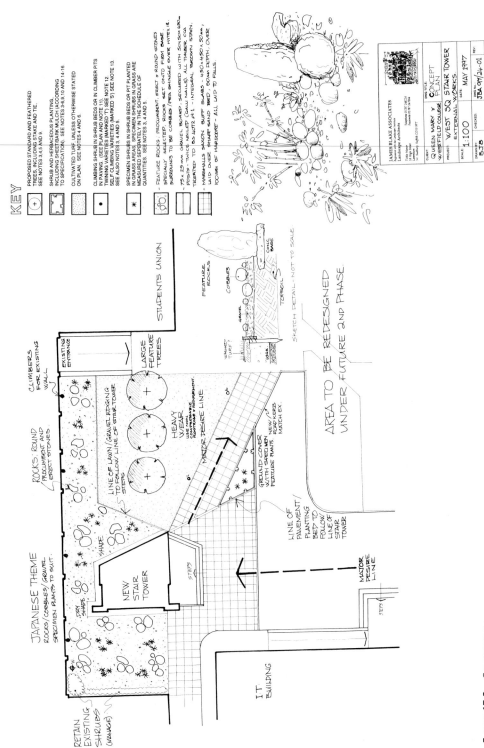

Figure 17.2 Concept plan

enhanced to produce a slickly drawn presentation plan, which sells the ideas to the client, the planners and the general public. The graphics will show many textures and pen strokes, with shading and shadows to present a three-dimensional effect. The plan will be precisely and accurately drawn. Individual graphic style can be developed to the full with the emphasis on expression of ideas and character. Symbols that look like the feature that they represent should be used rather than indicative symbols used on maps or working drawings.

Planting should be shown using a wide range of pen widths to reveal the different heights of plants intended. By altering the texture and pattern of the shrub symbols, you can illustrate the variety of foliage effects that you intend to create. Trees should be drawn with branches or leaves, but the outer ring should still be a fine pen, though the line here can be thicker on the north (shaded) side to help attain a three-dimensional effect to the drawing. Sciagraphy is essential. Shadows should be cast to the north or north-west at either 30 degrees or 45 degrees. The shadow should be shown both on the side of the object away from the sun and on the ground cast by the object. Ground textures should be shown, but not be too dominant – fine textures are better than coarse.

Final design layout drawings are excellent for public consultation meetings, public inquiries and for marketing your practice to potential clients. They look good framed in your office and clients may like them framed in their lounge or office. Final design layout drawings are especially powerful if accompanied with high quality perspective renditions, which can be in colour. Indeed watercolour paintings are very appropriate to landscape design presentation. Such perspective renditions can be mounted and keyed to the design layout plan. Such a drawing package can be very helpful where the design is open to public scrutiny and comment, as in the case of contentious development proposals at the planning and public consultation stage. Indeed public consultation is a fundamental part of public sector design projects such as parks and play areas.

Plans for a contractor

Working drawings need to be absolutely clear and totally precise, using a minimum of embellishment and conventional symbols that are instantly recognizable. If such drawings fulfil the above criteria, they will be of great assistance in realizing the design ideas in the form anticipated and without need of compromise.

The symbols used for such drawings are simple, clear, conventional (as far as is practical) and distinctive (from similar symbols). Use appropriate line weight as there will be few textures and graphic effects to describe the difference between surfaces.

Every item must be shown clearly and sufficiently. Choose an appropriate scale to enable you to show edgings, including 38 mm wide timber edgings. All edgings should be shown as a double line – to show that there is a thickness. If your design requires an edge restraint and you missed it off the plan, it is more than likely that you will not measure this item in the measured works

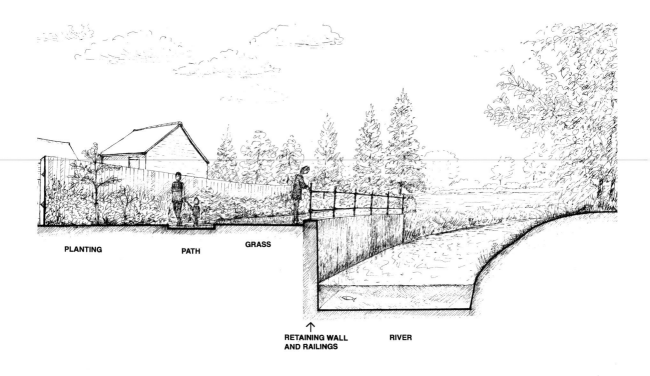

PLANTING   PATH   GRASS

↑
RETAINING WALL   RIVER
AND RAILINGS

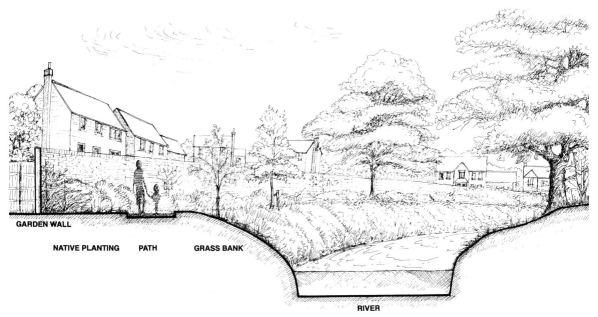

GARDEN WALL

NATIVE PLANTING   PATH   GRASS BANK

RIVER

FIGURE 17.3 RIVERSIDE
WALK – elevation
drawings

FIGURE 17.3 (concluded) RIVERSIDE WALK – perspective renditions

schedule and therefore it will not be priced and will have to be added on as an extra (at extra cost to the client). For annotation and the key, see the explanation on pp. 265–6.

The location plan

This plan will be exactly the same as that shown on the design drawings.

The hard works layout plan

This plan is normally simply referred to as the layout plan, as it shows everything except detailed planting information. The British Standard symbol for shrub planting is still used to outline where planting areas are to be positioned, but no division of the beds and no specific plants will be shown.

The plan is essentially devoted to the hard works elements of the schemes and all materials, levels, changes of level, manholes (and invert levels), gulleys, grilles, drainage and other service runs, kerbs, ramps, steps, edging, paving types, walls, fences, railings, gates, lighting and street furniture are shown, clearly labelled or itemized in the key and explained.

Where the information required to explain an element is too long-winded or complicated to write out in the key, then a number reference is placed by the item in a circle – referring to a component construction detail – and such references will be D1, D2, D3 and so on. The key might then simply say 'Supply and lay brick on flat paving in herring-bone bond – in accordance with Detail No. 1.' This states clearly that Detail 1 refers to herring-bone brick paving, and cannot be confused with an area of basket weave paving nearby, and that all detailed information can be found on the detail.

Where you need to show how different details actually relate and fit together, sometimes it is necessary to include a running section across the site. Such a section will be drawn on a separate sheet with a separate drawing number, unless there is spare space on the main plan. A section line will in any event be shown on the plan to show precisely where the section is drawn. This section line is drawn in a 0.6–0.8 pen and will normally not traverse the entire plan (which may obscure some vital feature) but will show just the beginning and the end of the line – usually 2.5 cm long each end – and then return for about 1 cm – in the direction in which the section is looking. At the ends of the returns an arrow will be drawn. Beside each arrow will be written a large capital letter, usually 'A' or 'B' etc. The detail section will then be referred to as Section A-A or Section B-B and so on. This is not to be confused with the often used annotation A–A^1, B–B^1, etc. which is used to describe the length of a line (depicting a fence, a wall, a hedge, etc.).

The planting plan

The planting plan is solely concerned with the soft works (trees, shrubs and grass) proposed. The proposed planting beds will be divided up to allow the designer to make detailed plant choices. This can be achieved in several different ways. One of the simplest methods to grasp is the circle method. This involves drawing circles to match the spacing of the individual plants chosen, drawing a small spot in the centre of each circle to represent the stem. These spots can then be joined using a 0.18 pen to show that they are all of the same species, and a pointer will then be taken to the edge of the plan and labelled – 4 No. *Choisya ternata*, for example – always starting with the quantity and followed by the plant name in Latin. The quantity is written '5 No.' rather than just '5' by convention (in all architectural specification), to avoid any confusion with catalogue reference numbers, names, etc.

The main drawback with the circle method is that it is only suitable for small detailed areas – drawn at 1:50 scale – both because of the longer time necessary to draught the plan and because of the complexity of the plan that emerges using the method.

The quickest and most common approach involves simply dividing the shrub bed concerned into a series of connecting boxes and then taking a pointer to the side of the plan and giving the quantity and species in the usual way. The quantity of plants is calculated by measuring the area of the box and then multiplying the area by the spacing suitable for the particular plant variety.

The setting out plan

The setting out plan will either be at the same scale as the hard works layout plan or at a larger scale, as this is the plan used literally to set out the items shown on the layout plan on to the ground. The plan will simply show the outline of every element with just enough detail to be able to distinguish between the lines, but with minimal embellishment. This plan is entirely for providing dimensions and measurements (levels may be provided but are often shown on the hard works layout plan only) and it must be as clear and uncluttered as possible (see Figures 17.4 and 17.5).

The distance between two edges of, say, a path will often be best shown by a dimension line (0.18 pen for the line) and though the line is drawn 20 mm or so beyond the edges of the path, the point at which the line crosses the path edge is marked with a diagonal line 3–4 mm long (using a 0.4 pen). The dimension figure will be written above the line centrally.

In order to define a sequence of points and lines across the site it is necessary to start with a known fixed point (such as the corner of an existing building) and then take a running measurement across the site. A line is drawn (0.18 pen) across the site and all planned elements that cross this line can be marked on by a 5 mm long perpendicular line (making a square cross) and then an arrow can be drawn (0.4 pen) touching the intersection and pointing away from the start point. By each of such intersections a measurement figure will be written, usually square to the paper or perpendicular to the running measurement line. These figures will be cumulative. For example, travelling across a front garden from the house they might be 2 m (path); 4 m (bed 2 m wide); 10 m (lawn 6 m wide); 12 m (bed 2 m wide).

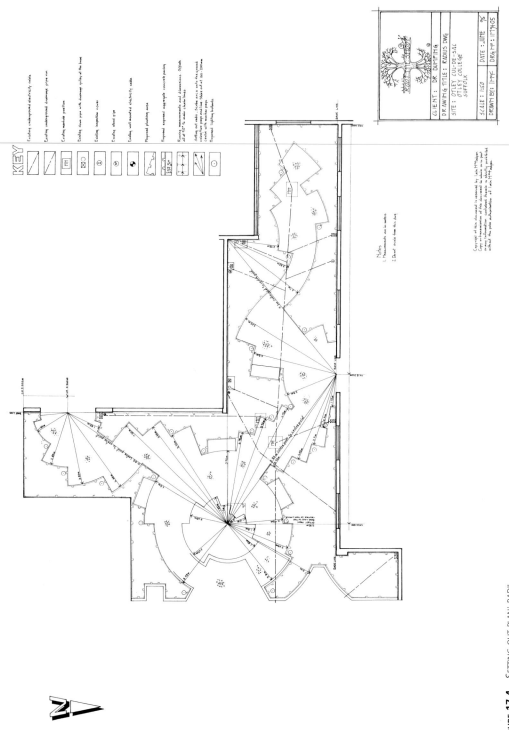

FIGURE 17.4   SETTING OUT PLAN: RADII

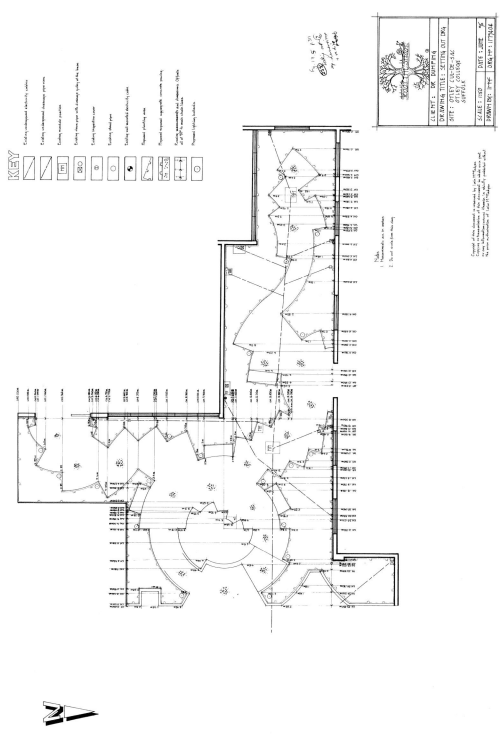

FIGURE 17.5   SETTING OUT PLAN: DIMENSIONS AND RUNNING MEASUREMENTS

When a point or line (of a planned element) is near to this running measurement line but does not cross it, you can still set out such points by taking an offset – either from your running measurement line or indeed from the site boundary. The running measurement line or boundary will be marked at the point opposite the element and the offset line taken at 90 degrees from it. An arrow is drawn against the element and the distance is then written against the arrow. Offsets can be used to set out any irregular shapes if spaced along a line at regular (say 2 m) intervals.

Regular curves should be shown using radius points, depicted with a cross. A line (0.18) is drawn from the cross to the edge of the circle and an arrow is drawn touching this perimeter line. The radius is written on the radius line centrally, for example R = 2.5 m. It is essential to fix the radius point by two dimensions from a boundary or other fixed points.

The demolition plan

This diagrammatic plan shows hatched and cross-hatched areas to be either demolished, excavated and carted away or retained, protected and made good. Much use is made of dotted, dashed and continuous lines to demarcate surfaces, structures, kerbs and edgings for retention or demolition.

The demolition plan is another plan that is drawn at a matching scale to the layout plan, that is 1:100 or 1:200. There are two main aspects to a demolition plan:

1. All items to be demolished, broken out, excavated or grubbed out/felled, etc. are highlighted with a texture or by hatching to define clearly the extent of each item. These items will be scheduled and a full description of the works required will be written on the plan.
2. All items to be retained but which require improvements or repairs are recorded on a separate schedule. Such reinstatement work is collectively known as 'dilapidations'.

Detailed plans

There are occasions where the majority of the site can be satisfactorily drawn at a scale of say 1:200, and the layout made perfectly clear to all contractors pricing or implementing the works. However, there may well be small areas of more complicated design and detailing which cannot be properly explained at the smaller scales, such as complex paving patterns or where levels, falls, drainage channels, kerbs and edgings are numerous and critical to the success of the scheme and to its use. Such areas require detail layout plans drawn at 1:50 scale, which can clearly show the component construction of all edgings and surfaces and how they intersect.

These plans are enlargements of parts of the main hard works layout plan and must be cross-referenced on both drawings. On the main hard works layout plan such areas will usually be encircled with a boundary line (usually a series of black spots or dashes) which will be labelled with reference to the plan number.

Such detail layout plans can be produced for the soft works elements too. It is particularly important to enlarge complicated ornamental or herbaceous planting schemes which require a high degree of precision and many different species closely juxtaposed. Once again 1:50 scale would be used, and the commonly used and fast 'block planting' method might be reluctantly forsaken for the slower but also surer circle method.

Construction details

Construction details are large-scale enlargements of particular elements of construction (usually hard works) that require additional explanation. The detail shows how the component parts fit together and so sections, elevations and plan views will be used. For a precast concrete (often abbreviated to 'pcc') slab pavement, a section will be needed to show the slabs, the sand bed and the hardcore base, all laid to falls over a compacted subgrade (or firm base of existing subsoil). The section may also show the edging treatment, which may be a brick on edge bedded on an in situ concrete foundation using cement mortar, and the soil level in relation to the pavement surface and edging level.

With paving a plan view, which will show the paving pattern, is often also drawn, usually on the same piece of paper. Precast slabs can be laid in a variety of ways in order and the pattern desired must be drawn and not left to chance.

While plans will almost always require pens of 0.18–0.2 to show paving patterns, plans of retaining walls will require a 0.4 pen to show the change of level. Sections are usually drawn using thin pens too, between the range 0.18 and 0.25, but all unit materials such as the bricks and slabs will usually be drawn in a 0.4 pen.

Dimension lines will be required to show the sizes of the components and thickness of bedding and base courses. This is necessary despite the fact that these construction details are drawn to scale. This is because the dimensions should be used by contractors rather than scaled off drawings – drawings may have been amended many times and the dimensions may get altered but the thicknesses often do not. Anyway the dimension is an accurate statement, whereas scaling drawings is never an exact science at the best of times. It is for this reason that most drawings will have the words 'do not scale from this drawing' printed on them.

Another common note on details is 'all dimensions are in millimetres' which saves having to write 'mm' after each figure. Consistency of dimension is also necessary, using all metres or millimetres as appropriate to the scale. Centimetres are never used for dimensioning on construction details by convention.

Elevations are not as often drawn when preparing construction details as an elevation portrays the feature (be it a wall, fence, etc.) above ground only. The main purpose of a construction detail is to show how the components fit together and most hard works items involve works below ground level, including foundations and footings (which must be drawn), and for these a section is essential. Elevations are mostly used for design drawings to show the appearance of a feature rather than its component construction. Having said that, for fencing the foundations may be adequately shown on an end section, and yet the face of the fence must be drawn to show how the rails and boards are joined to the posts, and here an elevation can be used. With fencing, a plan is sometimes necessary too, to show where the rails are joined to the posts.

Most construction details (99 per cent) are drawn at 1:20 or 1:10, though the range can be expanded to 1:20–1:5 to cover all eventualities and, just occasionally, full size drawings can be necessary.

CHAPTER 18

General management

THE main factors to consider when organizing the management of a professional landscape practice, in common with many design practices, can be summarized briefly as follows:

- The initial investment in the appropriate graphics equipment, computer equipment and furniture and ongoing investment in replacement items and improvements
- The marketing of the practice's design service and the establishment of a client base and a steady work flow
- The management of the production of landscape designs to meet client requirements and deadlines
- The management of work flow to ensure the minimum of delay in progress and to ensure a steady cash flow
- The recruitment of the right quality and quantity of personnel and the management of that personnel and their remuneration
- The efficient keeping of records, archives and accounts to comply with all statutory requirements and for general good business management and accountancy
- The continuing development of know-how and contemporary practices and the communication of such new methodology to all members of the practice
- Upholding the highest standards of professional practice and ensuring rigorous quality control to promote the practice's reputation and avoid expensive litigation.

Quality assurance

The efficient management of a professional landscape practice is essential for practitioners to fulfil their role and their duty of care to the client satisfactorily. If precision is the watchword of professionalism, then the effective control of commissions coming into the practice and efficient processing is clearly crucial to client satisfaction. Client satisfaction should be the primary goal of any business, in order to secure future commissions from that client and recommendations to new clients. While competitive fees are always at the forefront of most clients' minds both accuracy and diligence are arguably even more important.

Meeting deadlines, the recording and filing of information in a logical and retrievable manner and responding quickly to the client's needs and requests with the minimum inconvenience to the client are the key components of successful business and this is particularly applicable to landscape design practices. There are many ways that a business can be managed to achieve these aims but pre-conceived models for effective business organization do exist.

Successful business is an important goal both for the secure employment of the staff and for the country as a whole. One way to assist business to achieve success is to monitor its performance against an ideal template or blueprint and then to make improvements in order to redress any shortfalls in the business performance. This process can be applied to all aspects of the business, whether production, marketing, training and so on. The templates by which business operations and performance can be measured are collectively called a quality system. This system involves a standard way of doing things which can be used as a yardstick to gauge how a business is performing.

The international quality system BS EN ISO 9000 replaced an earlier UK system known as BS 5750 in 1994. The system is designed to be applicable to all business sectors. The benefits claimed are set out below:

- Improved productivity and efficiency, leading to cost savings
- Improved consistency of service/product performance and therefore greater customer satisfaction
- Improved customer perception of the organization's image, culture and performance
- Improved communications within the business leading to improved staff morale, job satisfaction and performance
- Improved competitive advantage and enhanced marketing opportunities.

The quality system demands certain criteria to be met. Managers must write a quality policy statement, explaining their goals and commitment to the quality system. A thorough review of what the customers want is required in order to ensure maximum customer satisfaction. This is called a contract review. The method of gearing resources to meet new customer requirements is termed design control, and servicing is the term given to aftersales service. Such procedures cover other areas of business operation such as:

- Purchasing
- Handling of products, people and property from the customer
- Process control (production)
- Handling storage and packaging
- Product identification and retrieval
- Training
- Inspection and testing
- Control of inspection, measuring and test equipment
- Inspection and test status
- Statistical techniques

- Control of non-conforming products – substandard quality products
- Document and data control
- Control of quality records
- Management responsibility and review
- Corrective and preventative action
- Internal quality audits.

Some companies who have adopted BS EN ISO 9000 have found that they have incurred time, expense and a considerable amount of bureaucracy without increasing their turnover, and this has resulted in a detrimental effect on profit. It is those firms that gain most of their work by reputation rather than from wider advertising that stand to benefit less from the marketing benefits of displaying conformity to a quality system.

Some local authorities insist on contractors acquiring BS EN ISO 9000 before they will allow a firm on to their approved lists of contractors. Therefore companies who work mainly for the public sector would have little choice but to join the scheme. The private sector, however, pays less attention to such regulation, caring more about efficiency and results themselves than the method of achieving them. For firms working mainly for the private sector managers must make a reasoned commercial judgement as to whether adopting the quality system is likely to improve their performance, turnover and profitability, or just tie up many staff hours in administrative hoop-jumping. There can be no doubt that there are some inefficient businesses and many more that could improve their procedures, and such a quality system might well be highly beneficial to them.

CHAPTER 19

The cost effectiveness of a guiding philosophy

█N conclusion, I hope that this fast track, grand tour of the subject of landscape architecture has shown clearly how the many different aspects fit together and has provided a good working knowledge of the main principles. I hope that, having achieved this understanding, the reader feels empowered to be able to achieve excellent, innovative and challenging design solutions and with the self-confidence to be able to ensure that these designs are implemented on the ground, to reflect the time, effort and creativity that have been put into the design.

It is hoped that the reader has been left with a rationale for the subject which will allow the assimilation of new information. The importance of establishing such a philosophy for the subject cannot be understated. Understanding the relevance of a piece of information to the many aspects of the subject is to be able to retrieve and use that information to the benefit of the design. If landscape designers are idealists then they are practical idealists; if landscape designers are philosophers then they are practical philosophers. As Bertrand Russell once said, 'There is nothing so practical for life as a good philosophy.' The landscape architect not only must be able to talk about ideas and innovations but must draw and describe them to the most finite detail and make them happen.

The purpose and role of industrial and professional bodies

All such bodies have a duty to regulate their respective industry or profession, to inform members and advise on training methods and courses, and to promote the value of the profession or trade to government, industry and to the public at large. In addition each body will fulfil the following roles:

- To advise its members and keep up to date with all new developments and to arbitrate in disputes

- To keep a register of members and a database of the work they carry out in order to advise potential clients which member would be able to provide a suitable service
- To provide a code of excellence and to provide incentives for high standards in the form of awards schemes.

Some of the professional bodies provide policy statements on good design practice, such as the Playing Fields Association's policy on the design of play areas, providing gradings for different age groups and sizes which should be provided according to the size of the residential development or community.

Often the most useful aspect of a professional body is in its advisory role and training role. The Landscape Institute maintains a register of members and its database of practices occasionally provides a commission.

Details of professional bodies that will be of interest to the landscape designer are provided below:

The Landscape Institute
Professional body for landscape architects, scientists and managers. 6/7 Barnard Mews, Clapham, London, SW11 1QU. Tel. (0171) 738 9166.

The Arboricultural Association
Professional association for arboriculturalists. Ampfield House, Romsey, Hampshire, SO51 9PA. Tel. (01794) 68717.

The British Association of Landscape Industries
Contractors trade association. Landscape House, 9 Henry Street, Keighley, West Yorkshire, BD21 3DR. Tel. (01535) 606139.

The British Standards Institute
389 Chiswick High Road
London W4 4AL. Tel. (01908) 220908.

The Concrete Advisory Service
37 Cowbridge Road
Pontyclun
Glamorgan CF72 9EB. Tel. (01443) 237210.

The Horticultural Trades Association
Trade association for the horticultural industry. Horticulture House, 19 High Street, Theale, Reading, Berkshire, RG7 5AH. Tel. (0118) 930 3132.

The Institute of Groundmanship
Professional body for park and playing field managers. 19–23 Church Street, The Agora, Wolverton, Milton Keynes, Buckinghamshire, MK12 5LG. Tel. (01908) 312511.

The Institute of Horticulture
Professional body for horticulturalists. PO Box 313, 80 Vincent Square, London, SW1P 2PE. Tel. (0171) 976 5951.

The Institute of Leisure and Amenity Management
Trade association for the leisure and sports management profession. Ilam House, Lower Basildon, Reading, Berkshire, RG8 9NE. Tel. (01491) 874222.

The National Playing Fields Association
25 Ovington Square
London SW3 1LQ. Tel. (0171) 584 6445.

The Sports Turf Research Institute.
St Ives Estate, Bingley, West Yorkshire, BD16 1AU. Tel. (01274) 565131.

CHAPTER 20

Further reading

Hugh Clamp (1999) *Landscape Professional Practice*, Aldershot: Gower.

Brian Davis (1987) *The Gardener's Illustrated Encyclopedia of Trees and Shrubs: A Guide to more than 2000 Varieties*, London: Viking.

C. A. Fortlage and E. T. Phillips *Landscape Construction. Vol. 1: Walls, Fences and Railings* (1992); *Vol. 2: Roads, Paving and Drainage* (1996), Aldershot: Gower.

Sir Geoffrey and Susan Jellicoe (1995) *The Landscapes of Man*, London: Thames and Hudson.

Adrian Lisney and Ken Fieldhouse (1990) *Landscape Design Guides. Vol. 1: Soft Landscape*; *Vol. 2: Hard Landscape*, Aldershot: Gower.

John Parker and Peter Bryan (1989) *Landscape Management and Maintenance: A Guide to its Costing and Organization*, Aldershot: Gower.

Ronald Fraser Reekie (1995) *Reekie Architectural Drawings* (Fourth Revision), Sevenoaks: Edward Arnold.

Nick Robinson (1992) *Planting Design Handbook*, Aldershot: Gower.

Construction detail

Figure A.1 shows the minimum requirement for the treatment of topsoil in connection with new development. Often the subsoil is compacted and needs breaking up to prevent waterlogging, and it is essential to ensure that topsoil levels are spread to 65 mm below any adjacent grass or paved surfaces to receive sheet mulch and bark mulch. This level should be reduced to 90 mm where sheet mulch is not specified in order that the bark mulch depth can be increased to 75 mm in order to be more effective in weed suppression. Sheet mulch is often given many names including aggrotextile, geotextile, and various brand names such as Mypex, Plantex, Weedex and Hy-tex (the latter being the most cost-effective to date in the opinion of the author).

Do not use sheet mulch unless the sub base works are carried out and the soil levels correct or it will not work. Also avoid using it on gradients greater than 1:4 and sites with heavy, sticky clay. Ensure that the sheet mulch is pegged down at 500 mm centres, closer where necessary and along edges. Use a no-fines coarse-grained woodchip to prevent the matter breaking down quickly and exposing sheet mulch.

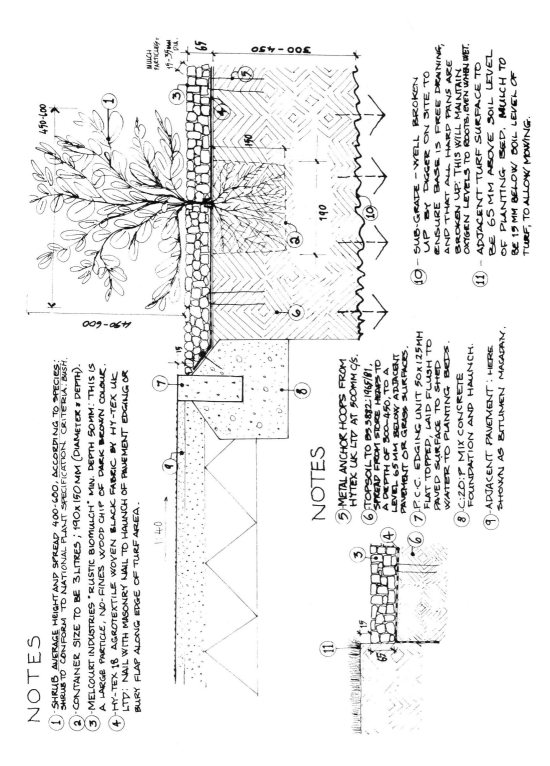

NOTES

1. SHRUB. AVERAGE HEIGHT AND SPREAD 400-600, ACCORDING TO SPECIES: SHRUBS TO CONFORM TO NATIONAL PLANT SPECIFICATION CRITERIA: BUSH.
2. CONTAINER SIZE TO BE 3 LITRES; 190 x 190 MM (DIAMETER & DEPTH).
3. MELCOURT INDUSTRIES "RUSTIC BIOMULCH" MIN. DEPTH 50MM: THIS IS A LARGE PARTICLE, NO-FINES WOOD CHIP OF DARK BROWN COLOUR.
4. HY-TEX 18 AGROTEXTILE WOVEN BLACK FABRIC BY HY-TEX UK LTD: NAIL WITH MASONRY NAIL TO HAUNCH OF PAVEMENT EDGING OR BURY FLAP ALONG EDGE OF TURF AREA.

NOTES

5. METAL ANCHOR HOOPS FROM HYTEX UK LTD AT 500MM C/S.
6. TOPSOIL TO BS 3882:1965/81, SPREAD FROM STORE HEAPS TO A DEPTH OF 300-450, TO A LEVEL 65 MM BELOW ADJACENT PAVEMENT OR GRASS SURFACES.
7. P.C.C. EDGING UNIT 50 x 125MM FLAT TOPPED, LAID FLUSH TO PAVED SURFACE TO SHED WATER TO PLANTING BEDS.
8. C:20:P MIX CONCRETE FOUNDATION AND HAUNCH.
9. ADJACENT PAVEMENT: HERE SHOWN AS BITUMEN MACADAM.

10. SUB-GRADE - WELL BROKEN UP BY DIGGER ON SITE TO ENSURE BASE IS FREE DRAINING, AND THAT ALL HARD PANS ARE BROKEN UP: THIS WILL MAINTAIN OXYGEN LEVELS TO ROOTS, EVEN WHEN WET.
11. ADJACENT TURF SURFACE TO BE 65 MM ABOVE SOIL LEVEL OF PLANTING BED, MULCH TO BE 15 MM BELOW SOIL LEVEL OF TURF, TO ALLOW MOWING.

FIGURE A.1   CONSTRUCTION DETAIL: SOIL AND MULCH

Index